Hacking Wireless Networks FOR DUMMIES®

by Kevin Beaver and Peter T. Davis

Foreword by Devin K. Akin
Chief Technology Officer,
The Certified Wireless Network Professional (CWNP) Program

WILEY

Wiley Publishing, Inc.

Hacking Wireless Networks For Dummies®

Published by
Wiley Publishing, Inc.
111 River Street
Hoboken, NJ 07030-5774

www.wiley.com

Copyright © 2005 by Wiley Publishing, Inc., Indianapolis, Indiana

Published by Wiley Publishing, Inc., Indianapolis, Indiana

Published simultaneously in Canada

For general information on our other products and services, please contact our Customer Care Department within the U.S. at 800-762-2974, outside the U.S. at 317-572-3993, or fax 317-572-4002.

For technical support, please visit www.wiley.com/techsupport.

Wiley also publishes its books in a variety of electronic formats. Some content that appears in print may not be available in electronic books.

Library of Congress Control Number: 2005924619

ISBN-13: 978-0-7645-9730-5

ISBN-10: 0-7645-9730-2

Manufactured in the United States of America

10 9 8 7 6 5 4 3 2 1

1O/ST/QY/QV/IN

WILEY

Hacking Wireless Networks For Dummies®

Cheat Sheet

Ethical Hacking Steps

This book covers dozens of steps, tricks, and methods for ethically hacking your wireless network. Here are some critical ones you can't afford to miss:

1. Plan your testing goals in detail.
2. Determine which tests to run in advance.
3. Get permission and signed authorization.
4. Gather your wireless-hacking tools.
5. Perform Internet searches, looking for "inside" information on your wireless systems.
6. Search for SSIDs — both clearly visible ones and hidden ones.
7. Warwalk/drive/fly to map out your wireless systems.
8. Look for unauthorized APs and ad-hoc wireless clients.
9. Check for common default settings in your APs and wireless clients.
10. Search for vulnerable network protocols and services.
11. Look for other odd protocol issues.
12. Attempt to bypass MAC-address access controls.
13. Take a crack at cracking your WEP keys.
14. Test EAP, LEAP, and other authentication schemes.
15. Analyze your results.
16. Follow up on — and plug — the security holes you find.

Wireless Standards

Protocol	Bandwidth	Frequency	Signaling Method	Distance Range	Included Security	Key Lengths
802.11a	54 Mbps	5-6 GHz	OFDM	50 to 90 feet (15 to 30 meters)	WEP	40-, 64- and 128-bit
802.11b	11 Mbps	2.4 GHz	DSSS	300 feet (100 meters)	WEP	40-, 64-, 128-, and 152-bit
802.11g	54 Mbps	2.4 GHz	DSSS, OFDM	150 to 300 feet (50 to 100 meters)	WPA, 802.1X, 802.11i (WPA2)	40-, 64-, 128-, 152-, 192-, and 256-bit

For Dummies: Bestselling Book Series for Beginners

Hacking Wireless Networks For Dummies®

Cheat Sheet

EAP Variants

- Lightweight EAP (LEAP) (http://www.cisco.com)
- EAP-TLS (http://www.microsoft.com or http://www.freebsd.org or http://www.linux.org)
- EAP-TTLS (http://www.funk.com or http://www.mtghouse.com)
- Protected EAP (PEAP) (http://www.microsoft.com)
- EAP-Subscriber Identity Module (SIM) (http://www.bizforum.org/whitepapers/intel-2.htm)
- EAP-MD5 (https://datatracker.ietf.org/public/idindex.cgi?command=id_detail&id=10146)

802.11a: Usable Channels

Frequency Band	Channel Number	Center Frequencies (GHz)
U-NII Lower Band (5.15 to 5.25 GHz)	36	5.180
	40	5.200
	44	5.220
	48	5.240
U-NII Middle Band (5.25 to 5.35 GHz)	52	5.260
	56	5.280
	60	5.300
	64	5.320
U-NII Upper Band (5.725 to 5.825 GHz)	149	5.745
	153	5.765
	157	5.785
	161	5.805

802.11b and 802.11g: Frequency and Channels

Channel	Frequency	Channel	Frequency
1	2.412 GHz	8	2.447 GHz
2	2.417 GHz	9	2.452 GHz
3	2.422 GHz	10	2.457 GHz
4	2.427 GHz	11[1]	2.462 GHz
5	2.432 GHz	12	2.467 GHz
6	2.437 GHz	13[2,3]	2.472 GHz
7	2.442 GHz	14[4]	2.477 GHz

1 North America uses channels 1-11

2 Europe (except France) uses channels 1-13

3 France uses channels 10-13

4 Japan uses channels 1-14

For Dummies: Bestselling Book Series for Beginners

About the Authors

Kevin Beaver is founder and information security advisor with Principle Logic, LLC, an Atlanta-based information-security services firm. He has over 17 years of experience in the IT industry and specializes in information security assessments for those who take security seriously — and incident response for those who don't. Before starting his own information-security services business, Kevin served in various information-technology and security roles for several healthcare, e-commerce, financial, and educational institutions.

Kevin is author of *Hacking For Dummies* as well as the e-book *The Definitive Guide to Email Management and Security* (Realtimepublishers.com). In addition, Kevin co-authored *The Practical Guide to HIPAA Privacy and Security Compliance* (Auerbach Publications). He was also a contributing author and editor of *Healthcare Information Systems*, 2nd ed., (Auerbach Publications), and technical editor of *Network Security For Dummies*.

Kevin is a regular columnist and information-security expert for SearchSecurity.com, SearchWindowsSecurity.com, SearchNetworking.com, SearchExchange.com, and SearchSmallBizIT.com. He also serves as a contributing editor for HCPro's Briefings on HIPAA newsletter and is a Security Clinic Expert for ITsecurity.com. In addition, Kevin's information-security work has been published in Information Security Magazine, SecurityFocus.com, and Computerworld.com. Kevin is an information-security instructor for the Southeast Cybercrime Institute, and frequently speaks on information security at various conferences for CSI, TechTarget, IIA, SecureWorld Expo, and the Cybercrime Summit.

Kevin earned his bachelor's degree in Computer Engineering Technology from Southern Polytechnic State University and his master's degree in Management of Technology from Georgia Tech. He also holds MCSE, Master CNE, and IT Project+ certifications. Kevin can be reached at kbeaver@principlelogic.com.

Peter T. Davis (CISA, CMA, CISSP, CWNA, CCNA, CMC, CISM) founded Peter Davis+Associates (a very original name) as a firm specializing in the security, audit, and control of information. A 30-year information-systems veteran, Mr. Davis's career includes positions as programmer, systems analyst, security administrator, security planner, information-systems auditor, and consultant. Peter is also the founder (and past President) of the Toronto ISSA chapter, past Recording Secretary of the ISSA's International Board, and past Computer Security Institute Advisory Committee member. Mr. Davis has written or co-written numerous articles and 10 books, including *Wireless Networks For Dummies* and *Securing and Controlling Cisco Routers*. In addition, Peter was the technical editor for *Hacking For Dummies* and *Norton Internet Security For Dummies*. Peter is listed in the *International Who's Who of Professionals*. In addition, he was only the third editor in the three-decade history of *EDPACS*, a publication in the field of security, audit, and control. He finds time to be a part-time lecturer in data communications at Seneca College (`http://cs.senecac.on.ca`). He lives with his wife Janet, daughter Kelly, two cats, and a dog in Toronto, Ontario.

Dedication

Little G — this one's for you. You're such a great motivator and inspiration to me — more than words can say. Thanks for reminding me of what's really important. Thanks for being you.

—*Kevin*

To all my friends and enemies. Hopefully, the first group is bigger than the second.

—*Peter*

Authors' Acknowledgments

Kevin:

Thanks to Melody Layne, our acquisitions editor, for approaching me about this project and getting the ball rolling.

I'd like to thank our project editor, Chris Morris, as well as Kevin Kirschner and all the behind-the-scenes copy editors for pulling this thing together. Many thanks to my co-author Peter T. Davis for working with me on this book. It has been an honor and a pleasure.

I'd also like to thank Hugh Pepper, our technical editor, for the feedback and insight he gave us during the technical editing process.

Also, many thanks to Devin Akin with Planet3 Wireless for writing the foreword. Major kudos too for all the positive things you've done for the industry with the CWNP program. You're a true wireless network pioneer.

Many thanks to Ronnie Holland with WildPackets, Chia Chee Kuan with AirMagnet, Michael Berg with TamoSoft, Matt Foster with BLADE Software, Ashish Mistry with AirDefense, and Wayne Burkan with Interlink Networks for helping out with my requests.

Thanks, appreciation, and lots of love to Mom and Dad for all the values and common sense you instilled in me long ago. I wouldn't be where I'm at today without it.

Finally, to my dear wife Amy for all her support during this book. Yet another one I couldn't have done without you! You're the best.

Peter:

Melody Layne (our acquisitions editor) for pitching the book to the editorial committee and getting us a contract. As always, much appreciated.

Chris Morris for helping us bring this project to fruition. Kudos, Chris.

Hugh Pepper, tech editor, for his diligence in reviewing the material. Thanks, Hugh, for stepping in and stepping up.

Peter would like to thank Kevin Beaver for suggesting we write this together. Thanks Kevin. Peter would also like to thank Ken Cutler, Gerry Grindler, Ronnie Holland, Carl Jackson, Ray Kaplan, Kevin Kobelsky, Carrie Liddie, Dexter Mills Jr. and Larry Simon for responding to a request for wireless information. Thanks for answering the call for help. And a really big shout-out to John Selmys and Danny Roy for their efforts. Thanks, guys. The provided information shows in this book. Peter would be remiss should he not thank the NHL and NHLPA for canceling the hockey season. Thanks for freeing up his time to write this book. But the book is done, so get it together so he has something to watch this fall! (Come on guys, the Raptors don't quite fill the void.) A special thanks to Janet and Kelly for allowing Peter to work on the book as they painted the family room. Now he can kick back and enjoy the room!

Publisher's Acknowledgments

We're proud of this book; please send us your comments through our online registration form located at www.dummies.com/register/.

Some of the people who helped bring this book to market include the following:

Acquisitions, Editorial, and Media Development

Project Editor: Christopher Morris

Acquisitions Editor: Melody Layne

Copy Editors: Barry Childs-Helton, Andy Hollandbeck, Beth Taylor

Technical Editor: Hugh Pepper

Editorial Manager: Kevin Kirschner

Editorial Assistant: Amanda Foxworth

Cartoons: Rich Tennant (www.the5thwave.com)

Composition Services

Project Coordinator: Adrienne Martinez

Layout and Graphics: Carl Byers, Andrea Dahl, Mary Gillot Virgin

Proofreaders: Jessica Kramer, Joe Niesen, Carl William Pierce, Dwight Ramsey, TECHBOOKS Production Services

Indexer: TECHBOOKS Production Services

Publishing and Editorial for Technology Dummies

Richard Swadley, Vice President and Executive Group Publisher

Andy Cummings, Vice President and Publisher

Mary Bednarek, Executive Acquisitions Director

Mary C. Corder, Editorial Director

Publishing for Consumer Dummies

Diane Graves Steele, Vice President and Publisher

Joyce Pepple, Acquisitions Director

Composition Services

Gerry Fahey, Vice President of Production Services

Debbie Stailey, Director of Composition Services

Contents at a Glance

Table of Contents

Foreword

In all of networking history, it has never been easier to penetrate a net-work. IEEE 802.11 wireless LAN technology gives the hacker and network-security professional inexpensive — many times free — tools to work with. Whether you are an avid user of Linux or Windows, the tools are everywhere. Due to the enduring and ubiquitous warez community, hackers can obtain even the expensive analysis and penetration tools — such as 802.11-protocol analyzers — with no investment.

This book will show you quite a few of the latest tools, but an exhaustive text covering all currently-available wireless hacking tools would require a forklift to move, and would require you to remove all other books from your book-shelves to make room. With this many available tools, the important factor becomes learning how to use them effectively and efficiently.

Beginners have wasted many weekends wardriving neighborhoods or busi-ness districts. This type of probing for low-hanging fruit yields little, and is a waste of valuable learning time. It is much more to an individual's benefit to learn an assortment of wireless-LAN penetration tools and work toward the goal of obtaining *useful* information. Learning the tools and techniques takes time and hard work in a closed environment, but yields much in the information-technology arena.

The current demand for wireless-security professionals is staggering. Those individuals who have taken the time to hone their skills in the use of available tools and the latest penetration techniques will be financially rewarded with a great career. I urge you to consider practicing and studying rather than dri-ving around from neighborhood to neighborhood hoping to send an e-mail through someone's cable modem.

One of the biggest problems with wireless networks today is the lack of intru-sion detection. Banks, investment firms, hospitals, law offices, and other orga-nizations that house sensitive information may have a corporate policy stating that wireless LANs are not allowed. They may think that this "no-use" policy keeps their networks safe and secure, but they are gravely mistaken. A rogue access point could be placed on their network by intruders or by employees, and without a wireless-intrusion detection system, there would be no way to know that all of their security mechanisms have been bypassed — giving full access to anyone within 300 feet of the facility. Wireless-security professionals should be able to use available tools to locate wireless LANs, disable unautho-rized access points, and test for a full array of wireless vulnerabilities.

One of the most difficult tasks for a consultant today is teaching customers about wireless LAN technology. Often, organizations understand neither the technology nor the risks associated with it. 802.11 networks have a significant ROI for some organizations, but inherently create a security hole so big that you could drive a truck through it. Organizations should carefully consider whether 802.11 networks are feasible and can be cost-justified. Many things go into the securing of 802.11 networks, from secure installation to end-user and IT staff training.

Forgetting to cover a single base in wireless-LAN security can lead to intrusion and financial disaster. The risks can often far outweigh the gain of using 802.11 technology, so organizations decide to have a no-use policy. Still, those organizations *must* consider how to protect from wireless intrusion. One of the tricks to getting customers to "bite" — commit to the notion of protecting their wireless LAN — is to give them a quick demonstration of hacking tools. If they have (for example) a heavily loaded 802.11g network secured with WEP, cracking their WEP key should open their eyes very quickly.

Keep in mind that these demonstrations should ALWAYS be done with the permission of a person in authority at the client organization — and in a closed environment. Doing otherwise can lead to criminal prosecution, defamation of your organization, and a plethora of other undesirable results.

Time is never the IT professional's friend. Staying abreast of the latest tools and techniques takes lots of hard work and time. Reading a book like this one is a worthy endeavor toward becoming an experienced wireless security professional.

I am a firm believer in picking a field of study and becoming the best you can be in that particular area. Wireless LAN technology is so deep and wide that it can easily consume all of your time, so focusing on being a wireless LAN *security* professional is a reasonable and attainable choice. The market demand, the pay, and the career itself are all good. Best wishes to all who choose this career path — or endeavor to increase their networking knowledge by reading great books like this one.

Devin K. Akin

Chief Technology Officer, The Certified Wireless Network Professional (CWNP) Program http://www.cwnp.com

Introduction

*W*elcome to *Hacking Wireless Networks For Dummies.* This book outlines plain-English, wireless-network hacker tricks and techniques you can use to ethically hack 802.11-based wireless networks (yours or someone else's if you've been given permission) and discover security vulnerabilities. By turning the tables and using ethical hacking techniques, you then have a leg up on the malicious hackers — you'll be aware of any vulnerabilities that exist and be able to plug the holes before the bad guys have a chance to exploit them.

When we refer to *ethical hacking,* we mean the professional, aboveboard, and legal type of security testing that you — as an IT professional — can perform as part of your job. Villains need not apply.

Wireless networks are popping up everywhere. They provide a lot of freedom but not without cost: All too many wireless networks are left wide open for attack. As with any other computer or network, you must be up on the latest security concepts to properly secure 802.11-based wireless networks. But locking them down involves more than just port-scanning testing and patching vulnerabilities. You must also have the right security tools, use the proper testing techniques, and possess a watchful eye. And *know your enemy:* It's critical to think like a hacker to get a true sense of how secure your information really is.

Ethical hacking is a means of using the bad-guy (black-hat) techniques for good-guy (white-hat) purposes. It's testing your information systems with the goal of making them more secure — and keeping them that way. This type of security testing is sometimes called *penetration testing, white-hat hacking,* or *vulnerability testing,* but it goes further than that as you'll see when we outline the methodology in this book.

If you use the resources provided in this book, maintain a security-focused mindset, and dedicate some time for testing, we believe you'll be well on your way to finding the weaknesses in your wireless systems and implementing countermeasures to keep the bad guys off your airwaves and out of your business.

The ethical hacking tests and system-hardening tips outlined in this book can help you test and protect your wireless networks at places like warehouses, coffee shops, your office building, your customer sites, and even at your house.

Who Should Read This Book?

If you want to find out how to maliciously break into wireless networks this book is not for you. In fact, we feel so strongly about this, we provide the following disclaimer.

If you choose to use the information in this book to maliciously hack or break into wireless systems in an unauthorized fashion — you're on your own. Neither Kevin nor Peter as the co-authors nor anyone else associated with this book shall be liable or responsible for any unethical or criminal choices you may make using the methodologies and tools we describe. This book and its contents are intended solely for IT professionals who wish to test the security of wireless networks in an authorized fashion.

So, anyway, this book is for you if you're a network administrator, information-security manager, security consultant, wireless-network installer, or anyone interested in finding out more about testing 802.11-based wireless networks in order to make them more secure — whether it's your own wireless network or that of a client that you've been given permission to test.

About This Book

Hacking Wireless Networks For Dummies is inspired by the original *Hacking For Dummies* book that Kevin authored and Peter performed the technical editing. *Hacking For Dummies* covered a broad range of security testing topics, but this book focuses specifically on 802.11-based wireless networks. The techniques we outline are based on information-security best practices, as well as various unwritten rules of engagement. This book covers the entire ethical-hacking process, from establishing your plan to carrying out the tests to following up and implementing countermeasures to ensure your wireless systems are secure.

There are literally hundreds, if not thousands, of ways to hack wireless network systems such as (for openers) laptops and access points (APs). Rather than cover every possible vulnerability that may rear its head in your wireless network, we're going to cover just the ones you should be most concerned about. The tools and techniques we describe in this book can help you secure wireless networks at home, in small-to-medium sized businesses (SMBs) including coffee shops, and even across large enterprise networks.

How to Use This Book

This book bases its approach on three standard ingredients of ethical-hacking wisdom:

✔ Descriptions of various non-technical and technical hack attacks — and their detailed methodologies

✔ Access information to help you get hold of common freeware, open-source, and commercial security-testing tools

✔ Countermeasures to protect wireless networks against attacks

Each chapter is as an individual reference on a specific ethical-hacking subject. You can refer to individual chapters that pertain to the type of testing you wish to perform, or you can read the book straight through.

Before you start testing your wireless systems, it's important to familiarize yourself with the information in Part I so you're prepared for the tasks at hand. You've undoubtedly heard the saying, "If you fail to plan, you plan to fail." Well, it applies especially to what we're covering here.

Foolish Assumptions

Right off the bat, we make a few assumptions about you, the IT professional:

✔ You're familiar with basic computer-, network-, wireless- and information-security-related concepts and terms.

✔ You have a wireless network to test that includes two wireless clients at a minimum but will likely include AP(s), wireless router(s), and more.

✔ You have a basic understanding of what hackers do.

✔ You have access to a computer and a wireless network on which to perform your tests.

✔ You have access to the Internet in order to obtain the various tools used in the ethical-hacking process.

✔ Finally, perhaps the most important assumption is that you've obtained permission to perform the hacking techniques contained in this book. If you haven't, make sure you do — *before* you do anything we describe here.

How This Book Is Organized

This book is organized into five parts — three standard chapter parts, a Part of Tens, and a part with appendixes. These parts are modular, so you can jump around from one part to another to your heart's content.

Part I: Building the Foundation for Testing Wireless Networks

In Chapter 1, we talk about why you need to be concerned with wireless security — and outline various dangers that wireless networks face. We also talk about various wireless-testing tools, as well as hacks you can perform. Chapter 2 talks about planning your ethical-hacking journey, and Chapter 3 talks about the specific methods you can use to perform your tests. Chapter 4 finishes things off by outlining various testing tools you'll need to hack your wireless systems.

Part II: Getting Rolling with Common Wi-Fi Hacks

This part begins with Chapter 5, in which we talk about various non-technical, people-related attacks, such as a lack of security awareness, installing systems with default settings, and social engineering. Chapter 6 talks about various physical security ailments that can leave your network open to attack. Chapter 7 covers common vulnerabilities found in wireless-client systems associated with wireless PC Cards, operating system weaknesses, and personal firewalls — any of which can make or break the security of your wireless network. In Chapter 8, we dig a little deeper into the "people problems" covered in Chapter 5 — in particular, what can happen when people don't *change the default settings* (arrgh). We talk about SSIDs, passwords, IP addresses, and more, so be sure to check out this vital information on an often-overlooked wireless weakness. In Chapter 9, we cover the basics of war driving including how to use stumbling software and a GPS system to map out your wireless network. We'll not only cover the tools and techniques, but also what you can do about it — and that includes doing it ethically before somebody does it maliciously.

Part III: Advanced Wi-Fi Hacks

In Chapter 10, we continue our coverage on war driving and introduce you to some more advanced hacking tools, techniques, and countermeasures. In Chapter 11, we go into some depth about unapproved wireless devices — we lay out why they're an issue, and talk about the various technical problems associated with rogue wireless systems on your network. We show you tests you can run and give you tips on how you can prevent random systems from jeopardizing your airwaves. In Chapter 12, we look at the various ways your communications and network protocols can cause problems — whether that's with MAC address spoofing, Simple Network Management Protocol (SNMP) weaknesses, man-in-the-middle vulnerabilities, and Address

Resolution Protocol (ARP) poisoning. In Chapter 13, we cover denial-of-service attacks including jamming, disassociation, and deauthentication attacks that can be performed against wireless networks and how to defend against them. In Chapter 14, you get a handle on how to crack WEP encryption; Chapter 15 outlines various attacks against wireless-network authentication systems. In these chapters, we not only show you how to test your wireless systems for these vulnerabilities but also make suggestions to help you secure your systems from these attacks.

Part IV: The Part of Tens

This part contains tips to help ensure the success of your ethical-hacking program. You find out our listing of ten wireless-hacking tools. In addition, we include the top ten wireless-security testing mistakes, along with ten tips on following up after you're done testing. Our aim is to help ensure the ongoing security of your wireless systems and the continuing success of your ethical hacking program.

Part V: Appendixes

This part includes an appendix that covers ethical wireless-network hacking resources and a glossary of acronyms.

Icons Used in This Book

This icon points out technical information that is (although interesting) not absolutely vital to your understanding of the topic being discussed. Yet.

This icon points out information that is worth committing to memory.

This icon points out information that could have a negative impact on your ethical hacking efforts — so pay close attention.

This icon refers to advice that can help highlight or clarify an important point.

Where to Go from Here

The more you know about how the bad guys work, how your wireless networks are exposed to the world, and how to test your wireless systems for vulnerabilities, the more secure your information will be. This book provides a solid foundation for developing and maintaining a professional ethical-hacking program to keep your wireless systems in check.

Remember that there's no one best way to test your systems because everyone's network is different. If you practice regularly, you'll find a routine that works best for you. Don't forget to keep up with the latest hacker tricks and wireless-network vulnerabilities. That's the best way to hone your skills and stay on top of your game. Be ethical, be methodical, and be safe — happy hacking!

Part I
Building the Foundation for Testing Wireless Networks

In this part . . .

*W*elcome to the wireless frontier. A lot of enemies and potholes lurk along the journey of designing, installing, and securing IEEE 802.11-based networks — but the payoffs are great. Learning the concepts of wireless security is an eye-opening experience. After you get the basics down, you'll be the security wizard in your organization, and you'll know that all the information floating through thin air is being protected.

If you're new to ethical hacking, this is the place to begin. The chapters in this part get you started with information on what to do, how to do it, and what tools to use when you're hacking your own wireless systems. We not only talk about what to do, but also about something equally important: what *not* to do. This information will guide, entertain, and start you off in the right direction to make sure your ethical-hacking experiences are positive and effective.

Chapter 1

Introduction to Wireless Hacking

. . . .

In This Chapter

▶ Understanding the need to test your wireless systems

▶ Wireless vulnerabilities

▶ Thinking like a hacker

▶ Preparing for your ethical hacks

▶ Important security tests to carry out

▶ What to do when you're done testing

. . . .

*W*ireless local-area networks — often referred to as WLANs or Wi-Fi networks — are all the rage these days. People are installing them in their offices, hotels, coffee shops, and homes. Seeking to fulfill the wireless demands, Wi-Fi product vendors and service providers are popping up just about as fast as the dot-coms of the late 1990s. Wireless networks offer convenience, mobility, and can even be less expensive to implement than wired networks in many cases. Given the consumer demand, vendor solutions, and industry standards, wireless-network technology is real and is here to stay. But how safe is this technology?

Wireless networks are based on the Institute of Electrical and Electronics Engineers (IEEE) 802.11 set of standards for WLANs. In case you've ever wondered, the IEEE 802 standards got their name from the year and month this group was formed — February 1980. The ".11" that refers to the wireless LAN working group is simply a subset of the 802 group. There's a whole slew of industry groups involved with wireless networking, but the two main players are the IEEE 802.11 working group and the Wi-Fi Alliance.

Years ago, wireless networks were only a niche technology used for very specialized applications. These days, Wi-Fi systems have created a multibillion-dollar market and are being used in practically every industry — and in every size organization from small architectural firms to the local zoo. But with this increased exposure comes increased risk: The widespread use of wireless systems has helped make them a bigger target than the IEEE ever bargained for. (Some widely publicized flaws such as the Wired Equivalent Privacy (WEP) weaknesses in the 802.11 wireless-network protocol haven't helped things, either.) And, as Microsoft has demonstrated, the bigger and more popular you are, the more attacks you're going to receive.

With the convenience, cost savings, and productivity gains of wireless networks come a whole slew of security risks. These aren't the common security issues, such as spyware, weak passwords, and missing patches. Those weaknesses still exist; however, networking without wires introduces a whole new set of vulnerabilities from an entirely different perspective.

This brings us to the concept of ethical hacking. *Ethical hacking* — sometimes referred to as *white-hat hacking* — means the use of hacking to test and improve defenses against *un*ethical hackers. It's often compared to penetration testing and vulnerability testing, but it goes even deeper. Ethical hacking involves using the same tools and techniques the bad guys use, but it also involves extensive up-front planning, a group of specific tools, complex testing methodologies, and sufficient follow-up to fix any problems before the bad guys — the black- and gray-hat hackers — find and exploit them.

Understanding the various threats and vulnerabilities associated with 802.11-based wireless networks — and ethically hacking them to make them more secure — is what this book is all about. Please join in on the fun.

In this chapter, we'll take a look at common threats and vulnerabilities associated with wireless networks. We'll also introduce you to some essential wireless security tools and tests you should run in order to strengthen your airwaves.

Why You Need to Test Your Wireless Systems

Wireless networks have been notoriously insecure since the early days of the 802.11b standard of the late 1990s. Since the standard's inception, major 802.11 weaknesses, such as physical security weaknesses, encryption flaws, and authentication problems, have been discovered. Wireless attacks have been on the rise ever since. The problem has gotten so bad that two wireless security standards have emerged to help fight back at the attackers:

 ✔ **Wi-Fi Protected Access (WPA):** This standard, which was developed by the Wi-Fi Alliance, served as an interim fix to the well-known WEP vulnerabilities until the IEEE came out with the 802.11i standard.

 ✔ **IEEE 802.11i (referred to as WPA2):** This is the official IEEE standard, which incorporates the WPA fixes for WEP along with other encryption and authentication mechanisms to further secure wireless networks.

These standards have resolved many known security vulnerabilities of the 802.11a/b/g protocols. As with most security standards, the problem with these wireless security solutions is not that the solutions don't work — it's that many network administrators are resistant to change and don't fully implement them. Many administrators don't want to reconfigure their existing wireless systems and don't want to have to implement new security mechanisms for fear of making their networks more difficult to manage. These are legitimate concerns, but they leave many wireless networks vulnerable and waiting to be compromised.

Even after you have implemented WPA, WPA2, and the various other wireless protection techniques described in this book, your network may still be at risk. This can happen when (for example) employees install unsecured wireless access points or gateways on your network without you knowing about it. In our experience — even with all the wireless security standards and vendor solutions available — the majority of systems are still wide open to attack. Bottom line: Ethical hacking isn't a do-it-once-and-forget-it measure. It's like an antivirus upgrade — you have to do it again from time to time.

Knowing the dangers your systems face

Before we get too deep into the ethical-hacking process, it will help to define a couple of terms that we'll be using throughout this book. They are as follows:

- **Threat:** A *threat* is an indication of intent to cause disruption within an information system. Some examples of threat agents are hackers, disgruntled employees, and malicious software (malware) such as viruses or spyware that can wreak havoc on a wireless network.

- **Vulnerability:** A *vulnerability* is a weakness within an information system that can be exploited by a threat. Some examples are wireless networks not using encryption, weak passwords on wireless access points or APs (which is the central hub for a set of wireless computers), and an AP sending wireless signals outside the building. Wireless-network vulnerabilities are what we'll be seeking out in this book.

Beyond these basics, quite a few things can happen when a threat actually exploits the vulnerabilities of a various wireless network. This situation is called *risk.* Even when you think there's nothing going across your wireless network that a hacker would want — or you figure the likelihood of something bad happening is very low — there's still ample opportunity for trouble. Risks associated with vulnerable wireless networks include

- Full access to files being transmitted or even sitting on the server
- Stolen passwords
- Intercepted e-mails
- Back-door entry points into your wired network
- Denial-of-service attacks causing downtime and productivity losses
- Violations of state, federal, or international laws and regulations relating to privacy, corporate financial reporting, and more

✔ "Zombies" — A hacker using your system to attack other networks making you look like the bad guy

✔ Spamming — A spammer using your e-mail server or workstations to send out spam, spyware, viruses, and other nonsense e-mails

We could go on and on, but you get the idea. The risks on wireless networks are not much different from those on wired ones. Wireless risks just have a greater likelihood of occurring — that's because wireless networks normally have a larger number of vulnerabilities.

The really bad thing about all this is that without the right equipment and vigilant network monitoring, it can be impossible to detect someone hacking your airwaves — even from a couple of miles away! Wireless-network compromises can include a nosy neighbor using a frequency scanner to listen in on your cordless phone conversations — or nosy co-workers overhearing private boardroom conversations. Without the physical layer of protection we've grown so accustomed to with our wired networks, anything is possible.

Understanding the enemy

The wireless network's inherent vulnerabilities, in and of themselves, aren't necessarily bad. The true problem lies with all the malicious hackers out there just waiting to exploit these vulnerabilities and make your job — and life — more difficult. In order to better protect your systems, it helps to understand what you're up against — in effect, to think like a hacker. Although it may be impossible to achieve the same malicious mindset as the cyberpunks, you can at least see where they're coming from technically and how they work.

For starters, hackers are likely to attack systems that require the least amount of effort to break into. A prime target is an organization that has just one or two wireless APs. Our findings show that these smaller wireless networks help stack the odds in the hackers' favor, for several reasons:

✔ Smaller organizations are less likely to have a full-time network administrator keeping tabs on things.

✔ Small networks are also more likely to leave the default settings on their wireless devices unchanged, making them easier to crack into.

✔ Smaller networks are less likely to have any type of network monitoring, in-depth security controls such as WPA or WPA2, or a wireless intrusion-detection system (WIDS). These are exactly the sorts of things that smart hackers take into consideration.

However, small networks aren't the only vulnerable ones. There are various other weaknesses hackers can exploit in networks of all sizes, such as the following:

✔ The larger the wireless network, the easier it may be to crack Wired Equivalent Privacy (WEP) encryption keys. This is because larger networks likely receive more traffic, and an increased volume of packets to be captured thus leads to quicker WEP cracking times. We cover WEP in-depth in Chapter 14.

✔ Most network administrators don't have the time or interest in monitoring their networks for malicious behavior.

✔ Network snooping will be easier if there's a good place such as a crowded parking lot or deck to park and work without attracting attention.

✔ Most organizations use the omnidirectional antennae that come standard on APs — without even thinking about how these spread RF signals around outside the building.

✔ Because wireless networks are often an extension of a wired network, where there's an AP, there's likely a *wired* network behind it. Given this, there are often just as many treasures as the wireless network, if not more.

✔ Many organizations attempt to secure their wireless networks with routine security measures — say, disabling service-set-identifier (SSID) broadcasts (which basically broadcasts the name of the wireless network to any wireless device in range) and enabling media-access control (MAC) address filtering (which can limit the wireless hosts that can attach to your network) — without knowing that these controls are easily circumvented.

✔ SSIDs are often set to obvious company or department names that can give the intruders an idea which systems to attack first.

Throughout this book, we point out ways the bad guys work when they're carrying out specific hacks. The more cognizant you are of the hacker mindset, the deeper and broader your security testing will be — which leads to increased wireless security.

Many hackers don't necessarily want to steal your information or crash your systems. They often just want to prove to themselves and their buddies that they can break in. This likely creates a warm fuzzy feeling that makes them feel like they're contributing to society somehow. On the other hand, sometimes they attack simply to get under the administrator's skin. Sometimes they are seeking revenge. Hackers may want to use a system so they can attack other people's networks under disguise. Or maybe they're bored, and just want to see what information is flying through the airwaves, there for the taking.

The "high-end" *uber*hackers go where the money is — literally. These are the guys who break into online banks, e-commerce sites, and internal corporate databases for financial gain. What better way to break into these systems than through a vulnerable wireless network, making the real culprit harder to trace? One AP or vulnerable wireless client is all it takes to get the ball rolling.

For more in-depth insight into hackers — who they are, why they do it, and so on — check out Kevin's book *Hacking For Dummies* (Wiley) where he dedicated an entire chapter to this subject. Whatever the reasons are behind all of these hacker shenanigans, the fact is that your network, your information, and (heaven forbid) your job are at risk.

There's no such thing as absolute security on any network — wireless or not. It's basically impossible to be completely proactive in securing your systems since you cannot defend against an attack that hasn't already happened. Although you may not be able to prevent every type of attack, you can prepare, prepare, and prepare some more — to deal with attacks more effectively and minimize losses when they do occur.

Information security is like an arms race — the attacks and countermeasures are always one-upping each other. The good thing is that for every new attack, there will likely be a new defense developed. It's just a matter of timing. Even though we'll never be able to put an end to the predatory behavior of unethical cyber thugs, it's comforting to know that there are just as many ethical security professionals working hard every day to combat the threats.

Wireless-network complexities

In addition to the various security vulnerabilities we mentioned above, one of the biggest obstacles to secure wireless networks is their complexity. It's not enough to just install a firewall, set strong passwords, and have detailed access control settings. No, wireless networks are a completely different beast than their wired counterparts. These days, a plain old AP and wireless network interface card (NIC) might not seem too complex, but there's a lot going on behind the scenes.

The big issues revolve around the 802.11 protocol. This protocol doesn't just send and receive information with minimal management overhead (as does, say, plain old Ethernet). Rather, 802.11 is highly complex — it not only has to send and receive radio frequency (RF) signals that carry packets of network data, it also has to perform a raft of other functions such as

✔ Timing message packets to ensure client synchronization and help avoid data-transmission collisions

✔ Authenticating clients to make sure only authorized personnel connect to the network

✔ Encrypting data to enhance data privacy

✔ Checking data integrity to ensure that the data remains uncorrupted or unmodified

For a lot of great information on wireless-network fundamentals, check out the book that Peter co-authored — *Wireless Networks For Dummies.*

In addition to 802.11-protocol issues, there are also complexities associated with wireless-network design. Try these on for size:

- Placement of APs relative to existing network infrastructure devices, such as routers, firewalls, and switches
- What type of antennae to use and where to locate them
- How to adjust signal-power settings to prevent RF signals from leaking outside your building
- Keeping track of your wireless devices — such as APs, laptops, and personal digital assistants (PDAs)
- Knowing which device types are allowed on your network and which ones don't belong

These wireless-network complexities can lead to a multitude of security weaknesses that simply aren't present in traditional wired networks.

Getting Your Ducks in a Row

Before going down the ethical-hacking road, it's critical that you plan everything in advance. This includes:

- Obtaining permission to perform your tests from your boss, project sponsor, or client
- Outlining your testing goals
- Deciding what tests to run
- Grasping the ethical-hacking methodology (what tests to run, what to look for, how to follow-up, etc.) before you carry out your tests

For more on the ethical-hacking methodology, see Chapter 3.

All the up-front work and formal steps to follow may seem like a lot of hassle at first. However, we believe that if you're going to go to all the effort to perform ethical hacking on your wireless network as a true IT professional, do it right the first time around. It's the only way to go.

The law of sowing and reaping applies to the ethical-hacking planning phase. The more time and effort you put in up front, the more it pays off in the long run — you'll be better prepared, have the means to perform a more thorough wireless-security assessment, and (odds are) you'll end up with a more secure wireless network.

Planning everything in advance saves you a ton of time and work in the long-term; you won't regret it. Your boss or your client will be impressed to boot!

Gathering the Right Tools

Every job requires the right tools. Selecting and preparing the proper security testing tools is a critical component of the ethical-hacking process. If you're not prepared, you'll most likely spin your wheels and not get the desired results.

Just because a wireless hacking tool is designed to perform a certain test, that doesn't mean it will. You may have to tweak your settings or find another tool altogether. Also keep in mind that you sometimes have to take the output of your tools with a grain of salt. There's always the potential for *false positives* (showing there's a vulnerability when there's not) and even *false negatives* (showing there's no vulnerability when there *is*).

The following tools are some of our favorites for testing wireless networks and are essential for performing wireless hacking tests:

- Google — yep, this Web site is a great tool
- Laptop computer
- Global Positioning System (GPS) satellite receiver
- Network Stumbler network stumbling software
- AiroPeek network-analysis software
- QualysGuard vulnerability-assessment software
- WEPcrack encryption cracking software

Starting in Chapter 6, we get to work with these tools in more detail later on in this book, when we lay out specific wireless hacks.

You can't do without good security-testing tools, but no one of them is "the" silver bullet for finding and killing off all your wireless network's vulnerabilities. A trained eye and a good mix of tools is the best combination for finding the greatest number of weaknesses in your systems.

It's critical that you understand how to use your various tools for the specific tests you'll be running. This may include something as informal as playing around with the tools or something as formal as taking a training class. Don't worry, we'll show you how to work the basics when we walk you through specific tests in Chapters 5 through 16.

To Protect, You Must Inspect

After you get everything prepared, it's time to roll up your sleeves and get your hands dirty by performing various ethical hacks against your wireless network. There are dozens of security tests you can run to see just how vulnerable your wireless systems are to attack — and Chapters 5 through 16 of this book walk you through the most practical and important ones. The outcomes of these tests will show you what security holes can — or cannot — be fixed to make your wireless network more secure. Not to worry, we won't leave you hanging with a bunch of vulnerabilities to fix. We'll outline various countermeasures you can use to fix the weaknesses you find.

In the next few sections, we outline the various types of security attacks to establish the basis for the vulnerability tests you'll be running against your wireless network.

Non-technical attacks

These types of attacks exploit various human weaknesses, such as lack of awareness, carelessness, and being too trusting of strangers. There are also physical vulnerabilities that can give an attacker a leg up on firsthand access to your wireless devices. These are often the easiest types of vulnerabilities to take advantage of — and they can even happen to you if you're not careful. These attacks include

- Breaking into wireless devices that users installed on their own and left unsecured

- *Social engineering* attacks whereby a hacker poses as someone else and coaxes users into giving out too much information about your network

- Physically accessing APs, antennae, and other wireless infrastructure equipment to reconfigure it — or (worse) capture data off it

Network attacks

When it comes to the nitty-gritty bits and bytes, there are a lot of techniques the bad guys can use to break inside your wireless realm or at least leave it limping along in a nonworking state. Network-based attacks include

- Installing rogue wireless APs and "tricking" wireless clients into connecting to them
- Capturing data off the network from a distance by walking around, driving by, or flying overhead
- Attacking the networking transactions by spoofing MAC addresses (masquerading as a legitimate wireless user), setting up man-in-the-middle (inserting a wireless system between an AP and wireless client) attacks, and more
- Exploiting network protocols such as SNMP
- Performing denial-of-service (DoS) attacks
- Jamming RF signals

Software attacks

As if the security problems with the 802.11 protocol weren't enough, we now have to worry about the operating systems and applications on wireless-client machines being vulnerable to attack. Here are some examples of software attacks:

- Hacking the operating system and other applications on wireless-client machines
- Breaking in via default settings such as passwords and SSIDs that are easily determined
- Cracking WEP keys and tapping into the network's encryption system
- Gaining access by exploiting weak network-authentication systems

Chapter 2

The Wireless Hacking Process

We teach courses on ethical hacking — and when you're teaching, you need an outline. Our teaching outline always starts with the introduction to the ethical-hacking process that comprises most of this chapter. Inevitably, when the subject of an *ethical* hacking process comes up, the class participants visibly slump into their chairs, palpable disappointment written all over their faces. They cross their arms across their chests and shuffle their feet. Some even jump up and run from class to catch up on their phone calls. Why? Well, every class wants to jump right in and learn parlor tricks they can use to amaze their friends and boss. But that takes procedure and practice. Without a defined process, you may waste time doing nonessential steps while omitting crucial ones. So bear with us for a while; this background information may seem tedious, but it's important.

Obeying the Ten Commandments of Ethical Hacking

In his book *Hacking For Dummies* (Wiley), Kevin discussed the hacker genre and ethos. In Chapter 1, he enumerated the Ethical Hacking Commandments. In that book, Kevin listed three commandments. But (as with everything in networking) the list has grown to fill the available space. Now these commandments were not brought down from Mount Sinai, but thou shalt follow these commandments shouldst thou decide to become a believer in the doctrine of ethical hacking. The Ten Commandments are

1. Thou shalt set thy goals.

2. Thou shalt plan thy work, lest thou go off course.

3. Thou shalt obtain permission.

4. Thou shalt work ethically.

5. Thou shalt work diligently.

6. Thou shalt respect the privacy of others.

7. Thou shalt do no harm.

8. Thou shalt use a scientific process.

9. Thou shalt not covet thy neighbor's tools.

10. Thou shalt report all thy findings.

Thou shalt set thy goals

When Peter was a kid, he used to play a game at camp called Capture the Flag. The camp counselors would split all the campers into two teams: one with a red flag and one with a blue flag. The rules were simple: If you were on the blue team, then you tried to find the red flag that the red team had hidden and protected, and vice versa. Despite appearances, this game could get rough — on the order of, say, Australian Rules Football. It was single-minded: Capture the flag. This single-mindedness is similar to the goals of a *penetration test,* a security test with a defined goal that ends either when the goal is achieved or when time runs out. Getting access to a specific access point is not much different from capturing a flag: Your opponent has hidden it and is protecting it, and you're trying to circumvent the defenses. Penetration testing is Capture the Flag without the intense physical exercise.

How does ethical hacking relate to penetration testing? Ethical hacking is a form of penetration testing originally used as a marketing ploy but has come to mean a penetration test of all systems — where there is more than one goal.

In either case, you have a goal. Your evaluation of the security of a wireless network should seek answers to three basic questions:

- ✔ What can an intruder see on the target access points or networks?
- ✔ What can an intruder do with that information?
- ✔ Does anyone at the target notice the intruder's attempts — or successes?

You might set a simplistic goal, such as finding unauthorized wireless access points. Or you might set a goal that requires you to obtain information from a system on the wired network. Whatever you choose, you must articulate your goal and communicate it to your sponsors.

Involve others in your goal-setting. If you don't, you will find the planning process quite difficult. The goal determines the plan. To paraphrase the Cheshire Cat's response to Alice: "If you don't know where you are going, any path will take you there." Including stakeholders in the goal-setting process will build trust that will pay off in spades later on.

Thou shalt plan thy work, lest thou go off course

Few, if any of us, have an unlimited budget. We usually are bound by one or more constraints. Money, personnel or time may constrain you. Consequently, it is important for you to plan your testing.

With respect to your plan, you should do the following:

1. **Identify the networks you intend to test.**

2. **Specify the testing interval.**

3. **Specify the testing process.**

4. **Develop a plan and share it with all stakeholders.**

5. **Obtain approval of the plan.**

Share your plan. Socialize it with as many people as you can. Don't worry that lots of people will know that you are going to hack into the wireless network. If your organization is like most others, then it's unlikely they can combat the organizational inertia to do anything to block your efforts. It is important, though, to remember that you do want to do your testing under "normal" conditions.

Thou shalt obtain permission

When it comes to asking for permission, remember the case of the Internal Auditor who, when caught cashing a payroll check he didn't earn, replied, "I wasn't stealing. I was just testing the controls of the system." When doing ethical hacking, don't follow the old saw that "asking forgiveness is easier than asking for permission." Not asking for permission may land you in prison!

You must get your permission in writing. This permission may represent the only thing standing between you and an ill-fitting black-and-white-striped suit and a lengthy stay in the Heartbreak Hotel. You must ask for — and get — a

---

Aw, we were just having fun . . .

In December 2004, a Michigan man became the first person ever convicted of wardriving (the unauthorized snagging of confidential information via wireless access points, discussed in Chapters 9 and 10). Prosecutors presented evidence that he and his cronies had scanned for wireless networks, had found the wireless access point of a hardware chain store, had used that connection to enter the chain's central computer system, and had installed a program to capture credit-card information.

"get out of jail free" card. This card will state that you are authorized to perform a test according to the plan. It should also say that the organization will "stand behind you" in case you are criminally charged or sued. This means they will provide legal and organizational support as long as you stayed within the bounds of the original plan (see Commandment Two).

Thou shalt work ethically

The term *ethical* in this context means working professionally and with good conscience. You must do nothing that is not in the approved plan or that has been authorized after the approval of the plan.

As an ethical hacker, you are bound to confidentiality and non-disclosure of information you uncover, and that includes the security-testing results. You cannot divulge anything to individuals who do not "need-to-know." What you learn during your work is extremely sensitive — you must not openly share it.

Everything you do as an ethical hacker must be aboveboard, and must support the goals of the organization. You should notify the organization whenever you change the testing plan, change the source test venue, or detect high-risk conditions — *and* before you run any new high-risk or high-traffic tests, as well as when any testing problems occur.

You must also ensure you are compliant with your organization's governance and local laws. Do not perform an ethical hack when your policy expressly forbids it — or when the law does.

Thou shalt keep records

Major attributes of an ethical hacker are patience and thoroughness. Doing this work requires hours bent over a keyboard in a darkened room. You may have to do some off-hours work to achieve your goals, but you don't have to wear hacker gear and drink Red Bull. What you do have to do is keep plugging away until you reach your goal.

In the previous commandment we talked about acting professionally. One hallmark of professionalism is keeping adequate records to support your findings. When keeping paper or electronic notes, do the following:

- Log all work performed.
- Record all information directly into your log.
- Keep a duplicate of your log.
- Document — and date — every test.
- Keep factual records and record all work, even when you think you were not successful.

This record of your test design, outcome, and analysis is an important aspect of your work. Your records will allow you to compile the information needed for a written or oral report. You should take care in compiling your records. Be diligent in your work and your documentation.

Thou shalt respect the privacy of others

Treat the information you gather with the utmost respect. You must protect the secrecy of confidential or personal information. All information you obtain during your testing — for example, encryption keys or clear text passwords — must be kept private. Don't abuse your authority; use it responsibly. This means you won't (for example) snoop into confidential corporate records or private lives. Treat the information with the same care you would give to your own personal information.

Thou shalt do no harm

The prime directive for ethical hacking is, "Do no harm." Remember that the actions you take may have unplanned repercussions. It's easy to get caught up in the gratifying work of ethical hacking. You try something, and it works, so you keep going. Unfortunately, by doing this you may easily cause an outage of some sort, or trample on someone else's rights. Resist the urge to go too far — and stick to your original plan.

Also, you must understand the nature of your tools. Far too often, people jump in and start using the tools shown in this book without truly understanding the full implications of the tool. They do not understand that setting up a monkey-in-the-middle attack, for example, creates a denial of service. Relax, take a deep breath, set your goals, plan your work, select your tools, and (oh yeah) read the documentation.

Many of the tools we discuss here allow you to control the depth and breadth of the tests you perform. Remember this point when you want to run your tests on the wireless access point where your boss connects!

Thou shalt use a "scientific" process

By this commandment, we don't mean that you necessarily have to follow every single step of the scientific process, but rather that you adopt some of its principles in your work. Adopting a quasi-scientific process provides some structure and prevents undue chaos (of the sort that can result from a random-walk through your networks).

For our purposes, the scientific process has three steps:

1. Select a goal and develop your plan.

2. Test your networks and systems to address your goals.

3. Persuade your organization to acknowledge your work.

We address the first two steps in previous commandments, so let's look at the third step here. Your work should garner greater acceptance when you adopt an empirical method. An empirical method has the following attributes:

- ✓ **Set quantifiable goals:** The essence of selecting a goal (such as capturing the flag) is that you know when you've reached it. You either possess the flag or you don't. Pick a goal that you can quantify: associating with ten access points, broken encryption keys or a file from an internal server. Time-quantifiable goals, such as testing your systems to see how they stand up to three days of concerted attack, are also good.

- ✓ **Tests are consistent and repeatable:** If you scan your network twice and get different results each time, this is not consistent. You must provide an explanation for the inconsistency, or the test is invalid. If we repeat your test, will we get the same results? When a test is repeatable or replicable, you can conclude confidently that the same result will occur no matter how many times you replicate it.

- ✓ **Tests are valid beyond the "now" time frame:** When your results are true, your organization will receive your tests with more enthusiasm if you've addressed a persistent or permanent problem, rather than a temporary or transitory problem.

Thou shalt not covet thy neighbor's tools

No matter how many tools you may have, you will discover new ones. Wireless hacking tools are rife on the Internet — and more are coming out all the time. The temptation to grab them all is fierce. Take, for instance, "wardriving" tools.

Early on, your choices of software to use for this "fascinating hobby" were limited. You could download and use Network Stumbler, commonly called NetStumbler, on a Windows platform, or you could use Kismet on Linux. But these days, you have many more choices: Aerosol, Airosniff, Airscanner, APsniff, BSD-Airtools, dstumbler, Gwireless, iStumbler, KisMAC, MacStumbler, MiniStumbler, Mognet, PocketWarrior, pocketWiNc, THC-RUT, THC-Scan, THC-WarDrive, Radiate, WarLinux, Wellenreiter WiStumbler, and Wlandump, to name a few. And those are just the free ones. You also could purchase AirMagnet, Airopeek, Air Sniffer, AP Scanner, NetChaser, Sniff-em, Sniffer Wireless . . . Well you get the idea. Should you have unlimited time and budget, you could use all these tools. But we suggest you pick one tool and stick with it. (We give you a closer look at some from this list in Chapters 9 and 10.)

Thou shalt report all thy findings

Should the duration of your test extend beyond a week, you should provide weekly progress updates. People get nervous when they know someone is attempting to break into their networks or systems — and they don't hear from the people who've been authorized to do so.

You should plan to report any high-risk vulnerabilities discovered during testing as soon as they are found. These include

- ✔ discovered breaches
- ✔ vulnerabilities with known — and high — exploitation rates
- ✔ vulnerabilities that are exploitable for full, unmonitored, or untraceable access
- ✔ vulnerabilities that may put immediate lives at risk

You don't want someone to exploit a weakness that you knew about and intended to report. This will not make you popular with anyone.

Your report is one way for your organization to determine the completeness and veracity of your work. Your peers can review your method, your findings, your analysis, and your conclusions, and offer constructive criticism or suggestions for improvement.

If you find that your report is unjustly criticized, following the Ten Commandments of Ethical Hacking, should easily allow you to defend it.

One last thing: When you find 50 things, *report on* 50 things. You need not include all 50 findings in the summary but you must include them in the detailed narrative. Withholding such information conveys an impression of laziness, incompetence, or an attempted manipulation of test results. Don't do it.

Understanding Standards

Okay, we've told you that you need to develop a testing process — here's where we give you guidance on how to do so. We wouldn't keep you hanging by a wire (this is, after all, a wireless book). The following standards (which we get friendly with in the upcoming sections) provide guidance on performing your test:

- ISO 17799
- COBIT
- SSE-CMM
- ISSAF
- OSSTMM

You may find that the methodology you choose is preordained. For instance, when your organization uses COBIT, you should look to it for guidance. You don't need to use all of these methodologies. Pick one and use it. A good place to start is with the OSSTMM.

Using ISO 17799

The ISO/IEC 17799 is an internationally adopted "code of practice for information security management" from the International Organization for Standardization (ISO). The international standard is based on British Standard BS-799. You can find information about the standard at www.iso.org.

ISO/IEC 17799 is a framework or guideline for your ethical hack — not a true methodology — but you can use it to help you plan. The document does not specifically deal with wireless, but it does address network-access control. The document is a litany of best practices at a higher level than we would want for a framework for ethical hacking.

One requirement in the document is to control access to both internal and external networked services. To cover this objective, you need to try to connect to the wireless access point and try to access any resource on the wired network.

The document also requires that you ensure there are appropriate authentication mechanisms for users. You can test this by attempting to connect to a wireless access point (AP). When there is Open System authentication (see Chapter 16) you need not do any more work. Obviously no authentication is not appropriate authentication. APs with shared-key authentication may require you to use the tools shown in Chapter 15 to crack the key. If the AP is using WPA security, then you will need to use another tool, such as WPAcrack.

Should the AP implement Extensible Authentication Protocol (EAP), you may need a tool such as `asleap` (see Chapter 16).

Bottom line: These guidelines don't give you a step-by-step recipe for testing, but they can help you clarify the objectives for your test.

Using CoBIT

CoBIT is an IT governance framework. Like ISO 17799, this framework will not provide you with a testing methodology, but it will provide you with the objectives for your test.

You can find information about CoBIT at `www.itgi.org/`.

Using SSE-CMM

Ever heard of the CERT? (Give you a hint: It's not a breath mint or a candy.) It's the Computer Emergency Response Team that's part of the Software Engineering Institute (SEI) at Carnegie Mellon University in Pittsburgh, Pennsylvania. Well, the SEI is known for something else: It developed a number of *capability maturity models (CMM)* — essentially specs that can give you a handle on whether a particular system capability is up to snuff. The SEI included a CMM just for security — the Systems Security Engineering CMM (SSE-CMM for short). Now, the SSE-CMM won't lay out a detailed method of ethical hacking, but it can provide a framework that will steer you right. The SSE-CMM can help you develop a scorecard for your organization that can measure security effectiveness.

You can find out about SSE-CMM at `www.sei.cmu.edu/`.

The Computer Emergency Response team also sends out security alerts and advisories. The CERT has a methodology as well — OCTAVE. OCTAVE stands for Operationally Critical Threat, Asset, and Vulnerability Evaluation. You can use OCTAVE as a methodology to build a team, identify threats, quantify vulnerabilities, and develop an action plan to deal with them.

You can find OCTAVE at `www.cert.org/octave`.

Using ISSAF

The Open Information System Security Group (`www.oissg.org`) has published the Information Systems Security Assessment Framework (ISSAF). Developed as an initiative by information-security professionals, the ISSAF is a practical tool — a comprehensive framework you can use to assess how your security effectiveness. It's an excellent resource to use as you devise your test. (Draft 0.1 has, in fact, 23 pages on WLAN security assessment.)

The ISSAF details a process that includes the following steps:

1. Information gathering
    a. Scan
    b. Audit
2. Analysis and research
3. Exploit and attack
4. Reporting and presentation

These steps correspond to our Ten Commandments of Ethical Hacking. For each of the steps just given, the document identifies appropriate tasks and tools. For example, the scanning step lists the following tasks:

✔ Detect and identify the wireless network

✔ Test for channels and ESSID

✔ Test the beacon broadcast frame and recording of broadcast information

✔ Test for rogue access points from outside the facility

✔ IP address collection of access points and clients

✔ MAC address collection of access points and clients

✔ Detect and identify the wireless network

The document recommends you use programs such as Kismet, `nmap`, and `ethereal` as tools for Step 1.

You also will find information in the document on the software you can use and the equipment you will need to build or acquire to do your assessment of your organization's wireless-security posture.

The document we reviewed was a beta version, but it shows promise and is worth watching. You can find the ISSAF at `www.oissg.org/issaf`.

Using OSSTMM

We do recommend you take a long and hard look at the OSSTMM — the Open Source Security Testing Methodology Manual (`www.osstmm.org`). The Institute for Security and Open Methodologies (ISECOM), an open-source collaborative community, developed the OSSTMM's methods and goals much along the lines of the ISSAF: as a peer-review methodology. Now available as version 3.0, the OSSTMM has been available since January 2001 and is more mature than the ISSAF.

You'll find that the OSSTMM gathers the best practices, standard legal issues, and core ethical concerns of the global security-testing community — but this document also serves another purpose: consistent definition of terms. The document provides a glossary that helps sort out the nuances of vulnerability scanning, security scanning, penetration testing, risk assessment, security auditing, ethical hacking, and security hacking. The document also defines white-hat, gray-hat, and black-hat hackers, so that by their metaphorical hats ye shall know them. But even more importantly (from your viewpoint as an ethical-hacker-to-be), it provides testing methodologies for wireless security, distilled in the following bullets:

- **Posture review:** General review of best practices, the organization's industry regulations, the organization's business justifications, the organization's security policy, and the legal issues for the organization and the organization's regions for doing business.
- **Electromagnetic radiation (EMR) testing:** Testing of the electromagnetic radiation emitted from wireless devices.
- **802.11 wireless-networks testing:** Testing of access to 802.11 WLANs.
- **Bluetooth network testing:** Testing of Bluetooth ad-hoc networks.
- **Wireless-input-device testing:** Testing of wireless input devices, such as mice and keyboards.
- **Wireless-handheld testing:** Testing of handheld wireless devices, such as personal digital assistants and personal electronic devices.
- **Cordless-communications testing:** Testing of cordless communications communication devices, such as cellular technology.
- **Wireless-surveillance device testing:** Testing of wireless surveillance or monitoring devices, such as cameras and microphones.
- **Wireless-transaction device testing:** Testing of wireless-transaction devices, such as uplinks for cash registers and other point of sale devices in the retail industry.
- **RFID testing:** Testing of RFID (Radio Frequency Identifier) tags.
- **Infrared testing:** Testing of infrared communications communication devices.
- **Privacy review:** General privacy review of the legal and ethical storage, transmission, and control of data, based on employee and customer privacy.

Each step has associated tasks that provide more detail and specific tests. As well, each step has a table that outlines the expected results. For example, expected results for Step 3 include these:

- ✔ Verification of the organization's security policy and practices — and those of its users.

- ✔ Identification of the outermost physical edge of the wireless network.

- ✔ Identification of the logical boundaries of the wireless network.

- ✔ Enumeration of access points that lead into the network.

- ✔ Identification of the IP-range (and possibly DHCP-server) of the wireless network.

- ✔ Identification of the encryption methods used for data transfer.

- ✔ Identification of the authentication methods of exploitable "mobile units" (that is, the clients) and users.

- ✔ Verification of the configuration of all devices.

- ✔ Determination of the flaws in hardware or software that facilitate attacks.

Obviously, you need to cut and paste these tests according to your needs. For instance, should your organization not have infrared, then you would skip Step 11.

The OSSTMM is available from www.isecom.org/osstmm/.

With resources like these, you have a methodology — and everything you need to map out your plan. But rather than leave you hanging there, the rest of the book shows you how to work through a methodology. In Chapter 3, you develop a methodology for a review. In Chapter 4, you select your weapons of mass disruption. Chapters 6 through 16 show you how to use the tools to test your security posture. The only thing left after that is to evaluate your results. So . . .

Chapter 3

Implementing a Testing Methodology

*B*efore you start testing your wireless network for security vulnerabilities, it's important to have a formal, ordered methodology in place. Ethical hacking is more than just running a wireless-network analyzer and scanning willy-nilly for open ports. There are some formal procedures that should be incorporated into your testing to do it right and get the most of it — these, for example:

- ✔ Gathering public information such as domain names and IP addresses that can serve as a good starting point

- ✔ Mapping your network to get a general idea of the layout

- ✔ Scanning your systems to see which devices are active and communicating

- ✔ Determining what services are running

- ✔ Looking for specific vulnerabilities

- ✔ Penetrating the system to finish things off

The next few sections discuss these points in greater detail. By planning things out (in the ways we covered in Chapter 2) and having specific goals and methods — even checklists — you'll know where you're headed and you'll know when your testing is complete. It'll make your job a lot easier, and document the steps you take. Speaking of which . . .

Keep a log of what you do and when. Your logging can be more thorough if you take screen captures with a tool such as SnagIt (available at www.tech smith.com). Such visual samples come in handy when you know you won't be able to reproduce the same information on your screen again. Logging can be helpful if you need to go back and look at which tests you ran, figure out what worked and what didn't, or refer to special notes. It's also just the professional thing to do, and wise in case any problems crop up. For instance, you can show your boss or client what you did and when you did it if he needs to track a potential security intrusion.

When you're performing any type of information-security testing, it's important to work methodically and make sure you're running the right tests on the right systems. This will help ensure you find the greatest number of security weaknesses. A nice side benefit is that it can minimize sloppiness and help keep you from crashing your systems.

Ethically hacking your wireless network requires testing your systems with a pretty specific goal in mind: a more secure network (well, yeah). Getting there involves looking at your wireless systems from many perspectives — including that of a hacker outside your work area and that of a regular user inside your work area.

Be flexible in your methodology. You must adapt the rules of warfare to your enemy. No security-testing parameters are set in stone. The bad guys are using the latest tools, so you need those tools. They're breaking in from across the street, so you need to test from across the street. They're searching for network-configuration information from across the Internet — and you should, too. No two ethical hacks are alike; you may need to adjust your procedures if the need arises.

Determining What Others Know

Hackers start out by poking and prodding their victims' systems to find weaknesses — and you should, too. Look at your network from an outsider's perspective; find out what's available to just about anyone. This is even more important in the wireless realm because we don't have the added veil of physical security (as we do with our hard-wired systems).

What you should look for

Here's what you need to search for to get started with your wireless network testing:

- Radio-signal strength
- Specific SSIDs that are being broadcast
- IP addressing schemes
- Encryption such as WEP or VPN traffic
- Hardware makes and models
- Software versions

Footprinting: Gathering what's in the public eye

The first formal step in your ethical-hacking methodology is to perform a high-level network reconnaissance called *footprinting*. If you're performing a broader information-security assessment (the sort Kevin covered in *Hacking For Dummies*), you may want to search for things like employee names, patents and trademarks, or company files. You can gather this information through places such as your organization's Web site or the U.S. Patent and Trademark Office Web site (at www.uspto.gov), or by using a search engine such as Google (www.google.com). Because wireless networks are more infrastructure-based and localized, publicly available information might not be as prevalent as it is for your other network systems. It still pays to take a gander and see what's out there.

Searching with Google

An excellent tool to get started with this process is Google. In fact, Google is one of our favorite tools for performing security assessments in general. It's amazing what you can do with it. You can perform a ton of Web and newsgroup queries to search for information about your wireless systems. You can perform keyword searches and more detailed queries (using Google's advanced tools) to look for network configuration information and more that has been accidentally or intentionally made public on the Internet. This information can give a hacker a leg up on attacking your wireless systems. For example, using advanced queries in Google, you can search for:

- Word-processing documents
- Spreadsheets
- Presentations
- Network diagrams
- Network Stumbler mapping files
- Network-analysis packet files

Foundstone has a neat Google query tool that performs some advanced Google queries that you might not have thought up on your own at

```
www.foundstone.com/resources/freetools.htm
```

We show you how to put Google to work for you with various tests we'll outline in later chapters.

Searching Wi-Fi databases

The next area to search for information about your wireless systems is online Wi-Fi databases. These databases contain information such as SSIDs, MAC addresses, and more on wireless APs that have been discovered by curious outsiders. To get an idea what we're talking about, browse to the WiGLE database at

```
www.wigle.net/gps/gps/GPSDB/query/
```

and see if any of your APs are listed. (You'll have to register if you're a first-time visitor.) Once in, you can submit various search options as shown in Figure 3-1.

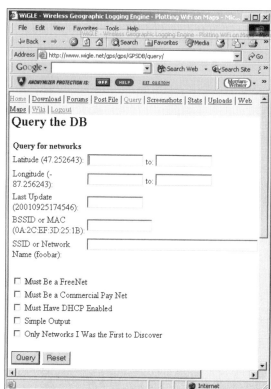

Figure 3-1:
Wireless-query options of the WiGLE database.

You can also check to see whether your AP is listed at another wireless lookup site at www.wifimaps.com.

You should also look up your domain name(s) at www.whois.org and your IP addresses at the American Registry for Internet Numbers (ARIN) site http://ws.arin.net/cgi-bin/whois.pl. These databases may be providing information about your wireless systems that you're not aware of or should not be advertising altogether.

Mapping Your Network

When you're satisfied that you've gotten a general view of what the general public can find out about your network without breathing hard, the next step is to create a network map to show how your wireless systems are laid out. You should do this from both inside and outside your network. That's necessary because wireless networks have a third dimension — the radio wave dimension — and tossing out all those radio waves allows them to be discovered from either side of your firewall or physical building. This allows you not only to see internal and external configuration information but also to see configuration information specific to wireless radio waves that are transmitted both "inside" and "outside" the network.

Compared to the way a typical wired network does its job, a wireless network opens a whole new dimension. Radio waves are like a virtual "third dimension" that can (in effect) allow hackers to jump over conventional boundaries.

Here are some suggestions for the best tools to help you map your network:

✔ **Network Stumbler:** The best tool to get started creating both internal and external maps of your wireless APs is Network Stumbler (www.netstumbler.com/downloads). This Windows-based tool allows you to scan the airwaves from outside your building to see what any hacker sitting in the parking lot or driving by can see. You can also run it from inside the confines of your building to look for any additional wireless APs that don't belong. Figure 3-2 shows the information that Network Stumbler can gather about your wireless network. (Note that the MAC addresses and SSIDs have been concealed to protect the innocent.)

✔ **AiroPeek:** A full-blown wireless network analyzer, or *sniffer,* AiroPeek (www.wildpackets.com/products/airopeek) is another great tool for gathering network map information. We demonstrate its features in later chapters.

Figure 3-2:
Network
Stumbler,
showing
various data
on APs
it has
discovered.

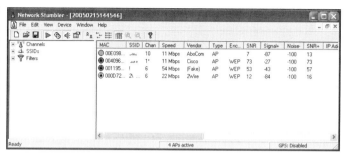

✔ **Cheops-ng and QualysGuard:** You'll want to gather not only information transmitted via RF but also any information about the wireless network that's only accessible via the internal wired network infrastructure. To do this bit of collecting, you can use a tool such as the open-source Cheops-ng (http://cheops-ng.sourceforge.net) or the commercial QualysGuard (www.qualys.com). Using either tool, you can create a network map of the inside of your network, which helps lay out the general IP addressing schemes and internal hostnames. You can also do this sort of mapmaking from outside your network to determine the names, external IP addresses, and registered domain name system (DNS) hostnames of publicly available hosts. Both methods give your wireless network more of a backbone, so to speak.

✔ **nmap and fping:** The other network mapping utilities in this list often utilize the Internet Control Message Protocol (ICMP) to determine which systems are "alive" on the network. Another way you can do this is by performing a ping sweep of your network, using a utility such as nmap (www.insecure.org/nmap) for Windows computers or a utility such as fping (www.fping.com) for UNIX and Linux. As shown in Figure 3-3, these tools won't create pretty graphical layouts of your network (that's the sort of thing you get from the network-mapping programs) but they're still very beneficial.

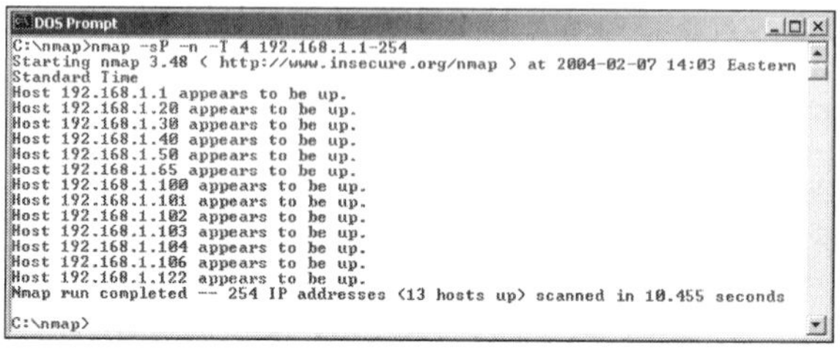

Figure 3-3:
nmap ping
sweep,
showing
which
systems are
alive on a
network.

On the outside looking in

It may seem tricky to try and scan your network from the "outside" — that is, from the public Internet. This is actually pretty simple. All you need is an available public IP address that you can assign to your test computer and plug in on the "public" side of your firewall or router that is connected directly to the Internet.

Note that the utilities we've mentioned in this section aren't necessarily going to be able to decipher which live systems are wireless and which ones are wired. It's up to you to determine which IP networks, IP addresses, and specific hostnames belong to your wireless devices. (You do have a recent network diagram, don't you? Just checking.)

Now you know which systems are alive on your network. The next step is to scan wireless systems for more information, such as open ports, hostnames, and more.

Scanning Your Systems

You've already gathered the higher-level information about your wireless systems such as SSIDs and IP addresses. You can find out more through a process called *enumeration*. Enumeration is when you examine a system and make an actual list of all the details you can discover about what it does and how. With enumeration, you can find

- ✔ Live wireless hosts (APs and ad-hoc clients)
- ✔ RF signal strength
- ✔ Whether WEP encryption is enabled
- ✔ Which network ports are open on wireless APs and clients

Network Stumbler, the friendly network-mapping tool mentioned a bit earlier, can not only find live wireless hosts (APs and ad-hoc clients), but it can also grab more in-depth information — for example, RF signal strength and whether WEP encryption is enabled. No surprise that Network Stumbler is a good tool for enumeration.

You can also get a little more in-depth goodies by using a port scanner such as nmap or SuperScan to poke and prod the network, looking to find out what network ports are open on your wireless APs and clients. You can find SuperScan at

```
www.foundstone.com/resources/proddesc/superscan.htm
```

Check your software licenses — even on free software — to make sure you're abiding by their restrictions. Many licenses state that the software cannot be used for commercial purposes. If you're doing your own internal testing, that might be okay, but testing wireless networks for paying clients is likely a no-no with this type of license limitation.

This port-scanning information helps create an even more detailed picture of what's available on your wireless network. No wonder hackers love it. This information gives them just what they need to try to exploit a ton of potential vulnerabilities on your systems. Table 3-1 outlines the ports that we often find open and vulnerable to attack, so be on special lookout for these.

Table 3-1	Commonly Hacked Wireless Network Ports	
Port Numbers	*Service*	*Protocols*
20	FTP data (File Transfer Protocol)	TCP
21	FTP control	TCP
22	SSH	TCP
23	Telnet	TCP
25	SMTP (Simple Mail Transfer Protocol)	TCP
53	DNS (Domain Name System)	UDP
80	HTTP (HyperText Transfer Protocol)	TCP
110	POP3 (Post Office Protocol version 3)	TCP
135	RPC/DCE end point mapper for Microsoft networks	TCP, UDP
137, 138, 139	NetBIOS over TCP/IP	TCP, UDP
161	SNMP (Simple Network Management Protocol)	TCP, UDP
443	HTTPS (HTTP over SSL)	TCP
512, 513, 514	Berkeley *r* commands (such as rsh, rexec, and rlogin)	TCP
1433	Microsoft SQL Server	TCP, UDP
1434	Microsoft SQL Monitor	TCP, UDP
3389	Windows Terminal Server	TCP

Determining More about What's Running

Performing port scans on your wireless network can snag a good amount of detail about how your system is set up and how it works — and knowledge is power. This power can be used to help you or used to harm you. But guess what? Once you know which ports are open, you (or somebody who isn't you) can find out even *more* details about the configuration of your wireless systems. Again, we're thinking like a hacker here — building a picture of what's available to be hacked.

By connecting to the open ports on your live systems, you can obtain even more in-depth enumeration information, such as:

- ✔ Acceptable usage policies and login warnings (or lack thereof) on banner pages
- ✔ Software and firmware versions (returned via banners or error messages)
- ✔ Operating-system versions (returned via banners, errors, or unique protocol fingerprints)
- ✔ Configurations of your operating system and applications

If by connecting to open ports you end up discovering a ton of exploitable information about your APs, workstations, and servers, don't panic. Many of your wireless systems may not be public-facing, so the likelihood of you having a lot of wireless devices with public IP addresses is low — at least it should be. However, you may have wireless hotspots or wireless-based servers that must be kept publicly accessible. These systems are usually reachable by an attacker — even through firewalls or other protective measures — and obtaining configuration information from them is not very difficult. This accessibility may entail some unavoidable vulnerability — a necessary evil if you want your network to do actual, useful work (what a concept).

Regardless, keep in mind that all this information can be used against you; eternal vigilance is the price of productivity. That leads us to performing an actual *vulnerability assessment* where you discover true vulnerabilities that can be exploited, which findings are false-positives, and which issues don't really matter.

Performing a Vulnerability Assessment

Now that you've found potential "windows" into your wireless network, the next step is to see whether any bigger vulnerabilities exist. In essence, you connect to the wireless systems and make a discreet, methodical attempt to see what can be found from a hacker's point of view. You may be able to gather more information, capture data out of thin air with a sniffer, or determine that a specific patch is missing.

Remember not to discount what you've already found simply because you're just now getting to the formal "vulnerability assessment" portion of the testing. Even without poking and prodding your wireless systems further, you may already have discovered some juicy vulnerabilities (such as default SSIDs, WEP not being enabled, and critical servers being accessible through the wireless network).

You can look for these vulnerabilities in two ways: manually and automatically. The next two sections discuss these two methods in greater detail.

Manual assessment

The first way — manual assessment — is the most time-consuming, but it's essential. Manually assessing vulnerabilities can be difficult at first, but it does get easier with experience. We call this assessment *manual,* but it often involves various semiautomatic security tools that don't just perform a run-of-the-mill robo-assessment but need your guiding hand now and then. Knowing how wireless networks and their associated operating systems and software work — knowing what's right and what stands out as a potential problem — can really help out your manual assessments. Be sure to check out Peter's book *Wireless Networks For Dummies* (as well as the other Linux and Windows *For Dummies* titles from Wiley) to learn more about these systems. Chey Cobb's book *Network Security For Dummies* is also a great resource to give you a handle on a wide variety of information-security concepts. Manual vulnerability-assessment techniques are a must — and we'll outline various ways to do them in various chapters throughout this book.

Automatic assessment

The second way of looking for vulnerabilities is to use an automated tool such as the open source Nessus (www.nessus.org) or the commercial LANguard Network Security Scanner (www.gfi.com/lannetscan) or QualysGuard. These tools can automate the vulnerability-assessment process by scanning live systems and determining whether vulnerabilities (actual or potential) exist. These tools take a lot of the legwork out of vulnerability assessment, giving you more time to spend catching up on e-mails or watching *Seinfeld* reruns. We can't imagine performing automated ethical-hacking tests without these tools.

There are many excellent wireless-hacking tools that you won't have to spend a dime on. They work great for running specific tests — but when it comes time to take a broader look at vulnerabilities, these commercial tools prove their worth. When you're delicately probing the operating systems and applications running on your wireless network, you definitely get what you pay for.

Finding more information

After you or your tools find suspected vulnerabilities, there are various wireless security vulnerability resources you can peruse to find out more information on the issues you find. A good place to start is your wireless vendor's Web site. Look in the Support or Knowledgebase section of the Web site for known problems and available security patches. You can also peruse the following vulnerability databases for in-depth details on specific vulnerabilities, how they can be exploited, and possible fixes:

  ✔ US-CERT Vulnerability Notes Database (`www.kb.cert.org/vuls`)

  ✔ NIST ICAT Metabase (`http://icat.nist.gov/icat.cfm`)

  ✔ Common Vulnerabilities and Exposures (`http://cve.mitre.org/cve`)

Another good way to get more information on specific security issues is to do a Google Web and groups search. Here you can often find other Web sites, message boards, and newsgroups where people have posted problems and (hey, let's be optimistic) solutions about your particular issue.

Penetrating the System

After you map your network, see which systems are running what, and find specific vulnerabilities, there's one more phase in your ethical-hacking methodology — *if* you choose to pursue it. This is the *system-penetration* phase: actually getting in. This is the true test of what systems and information can actually be compromised on your wireless network and the ultimate goal of malicious hackers.

Penetrating your wireless systems simply means acting as if you were completely unauthorized to access the resources on your wireless network — and trying to get in anyway. Sure, you've already been able to connect to your network "as an outsider" thus far, but this is where you take it all the way — by joining the wireless network, connecting to various systems, and doing things such as

✔ Logging in to the network

✔ Browsing the Internet

✔ Sending and receiving e-mails

✔ Changing AP configuration settings

✔ Capturing network data using a sniffer such as ethereal (www.ethereal.com) or AiroPeek

✔ Mapping to network drives

✔ Editing, copying, and deleting files — just be careful which ones!

The hackers are doing these things, so it may make sense to try them yourself so you can get a true view of what's possible on your network.

If you choose to penetrate your systems, proceed with caution — and work slowly and carefully to minimize any disruptions you may create!

In Parts II and III of this book, we outline how you can perform the tests we've described in this ethical-hacking methodology so you can start putting these procedures to work for you!

Chapter 4

Amassing Your War Chest

A cyberwar is being waged. Your perimeter is under siege. What makes the attack especially insidious is that you cannot see your enemy. This isn't hand-to-hand combat. Your enemy could be 2 miles from your office and still access your network and data. Your access point is your first line of defense in this war. It behooves you, then, to prepare for battle.

One way to prepare for any war is to participate in war games. Real war games allow you to test your equipment, tactics, and operations. In this case, war games allow you to test your wireless networks under normal conditions. Like the Reservist going off to war, you also must receive adequate training on the latest weapons and tactics. Although the rest of the book focuses on tactics, this chapter focuses on equipment. You need practice with the tools the crackers use for real.

You need some hardware and software, but you have choices about what type of hardware and software you use. This chapter serves as your armory. If you favor the Windows platform, we have some tools for you. Should you favor Linux, you will find some tools as well. We don't leave Apple enthusiasts out; we have something for you, too.

Choosing Your Hardware

What's your poison? Laptop or personal digital assistant? The two primary hardware platforms for wireless hacking are

- ✔ Personal digital assistant (PDA) or personal electronic device (PED)
- ✔ Portable or laptop

Each platform has its pros and cons. First, a PDA is readily portable so you can easily carry it from place to place. However, you won't find as many tools for the PDA as you will for other platforms — depending on the operating system you run on your handheld device. If you run the Zaurus operating system, for example, you have more choices for software than you do if you choose the Pocket PC operating system.

One thing is safe to say: You don't want to run wireless-hacking tools on a desktop. You may want to store NetStumbler files on the desktop, but the desktop is not really portable. The key thing to think about when choosing your hardware is *portability*. When performing hacking tests, you must be able to walk around your office building or campus, so a desktop is probably not the best choice. However, we know of people who use mini-towers in their cars for wardriving (discussed later in this chapter), but we don't recommend it!

The personal digital assistant

Because of its portability, a PDA is the perfect platform for wardriving — but not for tasks requiring processing power. You want to get a PDA that uses either the ARM, MIPS, or SH3 processor. We recommend the Hewlett-Packard iPAQ (ARM processor), the Hewlett-Packard Jornada (SH3 processor), or the Casio MIPS for wardriving. These are handy devices since someone was kind enough to develop network discovery software for these platforms.

ARM's processor technology has been licensed by more than 100 parties, so you should easily find a solution you like. It's so easy, in fact, that you would better spend your time choosing the right operating system for your needs. We tell you more about operating systems in the software section of this chapter.

The portable or laptop

PDAs are great, but, typically, ethical hackers use laptops. Laptops have dropped dramatically in price the last few years, so they have become more accessible. You don't need a lot of processing power, but, to paraphrase Tim

Allen, *more power is better.* You can use almost any operating system, including Windows 98, although you will find you get better results when using a newer and supported operating system. In addition to the laptop, you need the following components to get maximum results from your ethical hacking:

- ✔ Hacking software
- ✔ A wireless network interface card (NIC) that can be inserted into your laptop — preferably one with an external antenna jack
- ✔ External antenna (directional or omnidirectional) with the proper pigtail cable to connect your external antenna to your wireless NIC
- ✔ Portable global positioning system (GPS)
- ✔ DC power cable or DC to AC power inverter to power your laptop from your car's 12-volt DC cigarette lighter plug socket. These are widely available from RadioShack, Kmart, Staples, CompUSA, or Wal-Mart stores.

The next few sections discuss these components in greater detail.

Hacking Software

To do your job properly, you need a selection of freeware and commercial software. Fortunately, a glut of freeware programs is available, so you don't need a champagne budget; a beer budget should suffice. In fact, if you are prepared to run more than one operating system, you can get by using *only* freeware tools. You need the following software to do all the hacking exercises in this book:

- ✔ Partitioning or emulation software
- ✔ Signal strength–testing software
- ✔ Packet analyzer
- ✔ Wardriving software
- ✔ Password crackers
- ✔ Packet injectors

Using software emulators

In a perfect world, all the tools available would work on the same operating system. But in the real world, that's not the case. Many great tools operate on operating systems that are incompatible with each other. Very few of us, of course, are conversant with multiple operating systems. Also, few of us have the money to support duplicate hardware and software. So, how can you use all these tools? You need to find a solution that allows you to run more than one operating system on the same machine.

To solve this problem, people often build *dual-boot* or *multi-boot* workstations. You can use a product like Symantec's PartitionMagic (`www.symantec.com/partitionmagic`) to set up partitions for the various operating systems. For more information about setting up and using PartitionMagic, among other things, check out Kate Chase's *Norton All-in-One Desk Reference For Dummies* (Wiley). After you set up your partitions, you install the operating systems on the various partitions.

When everything's installed, you can select the operating system you want to use when you boot the system. Say you're using NetStumbler on Windows XP and you decide to use WEPcrack — which is available only on Linux — on the access points you just identified with NetStumbler. You shut down Windows XP, reboot your system, and select the Red Hat Linux operating system. When you want to use Windows XP again, you must do the reverse. This isn't a bad solution, but flipping back and forth a lot eats up valuable time. And managing your partitions and trying to make the operating systems coexist on the same hardware can be challenging.

Enter software emulators. *Software emulators* allow you to emulate a guest operating system by running it on top of a host operating system. You can run Linux emulation on a Windows host, and vice versa. To emulate Windows or DOS on a Linux host, you can choose one of the following Windows-based emulators:

- Bochs (`http://bochs.sourceforge.net`)
- DOSEMU (`www.dosemu.org`)
- Plex86 (`http://savannah.nongnu.org/projects/plex86`)
- VMware (`www.vmware.com`)
- WINE (`www.winehq.com`)
- Win4Lin (`www.netraverse.com`)

Alternatively, you can emulate Linux on a Windows host. To do this, choose one of the following Linux-based emulators:

- Cygwin (`http://cygwin.com`)
- VMware (`www.vmware.com`)

Mac lovers can already run most of the UNIX tools under the Mac OS. To emulate the Windows environment, you can run an emulator like Microsoft Virtual PC (www.microsoft.com/mac/products/virtualpc/virtualpc.aspx?pid=virtualpc).

To get you going, the next two sections discuss Cygwin and VMware, two excellent examples of emulation software.

Setting up Cygwin

Do you use Windows but have software that only runs on Linux? If so, Cygwin is your answer. Cygwin is a contraction of Cygnus + Windows. It provides a UNIX-like environment consisting of a Windows dynamically linked library (cygwin1.dll). Cygwin is a subsystem that runs on Windows and intercepts and translates UNIX commands. This is transparent to the user. With Cygwin, you can have the experience of running xterm and executing ls commands without ever leaving your safe Windows environment.

First, download Cygwin by going to http://cygwin.com. Installing Cygwin is easy when you follow these steps:

1. **On the home page, click the Install or Update Now! (Using setup.exe) link about halfway down the page.**

    You see a File Download – Security Warning window.

2. **Click Run to download Cygwin.**

    You see the message shown in Figure 4-1.

3. **Click Run to run setup.exe.**

    You see the Cygwin Setup window shown in Figure 4-2.

Figure 4-1:
Security
warning.

4. **Click Next.**

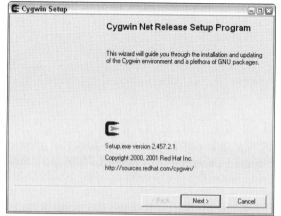

Figure 4-2:
Cygwin
setup.

5. Select Install from Internet and click Next.

This is the installation type. If you have a CD-ROM with Cygwin, select Install from Local Directory instead.

The Choose Installation Directory dialog box appears, as shown in Figure 4-3.

Figure 4-3:
Choosing
the
installation
directory.

6. Choose the installation directory options based on your needs and setup and then click Next.

After installation, this is the Cygwin root directory. Leave the default or click Browse to select another location. You can decide whether to make Cygwin available to all users or just to you. In addition, you can decide whether you want DOS or UNIX file types.

The ^M character

Ever see a ^M character at the end of your text files? Yes? Well, this likely means that someone transferred the file from UNIX to Windows using binary format rather than ASCII. DOS file lines end with a newline and a carriage return while UNIX file lines end with the newline only.

7. **Select a location in which to store the installation files. Then click Next.**

   Unless you have a compelling reason for not doing it, use the default. If you must put the installation files somewhere else, click Browse and select the location.

8. **Select the type of Internet connection you have. Click Next.**

   We suggest that when you aren't sure what to select here, use the default. If you're doing this from your home office, then Direct Connection should work. If you're at work, you might have a proxy server. If you have a proxy server, it's perhaps best to talk to your system administrator.

9. **Select a download site from the scroll box. Click Next.**

   You may have to try a few download sites before you find one that works for you. Peter tried several times to find a site. Either it would not start the download and required him to select a new site or it got halfway through the download and quit. You must persevere. Cygwin is worth it.

   The Select Packages window appears, as shown in Figure 4-4.

Figure 4-4: Selecting the packages to install.

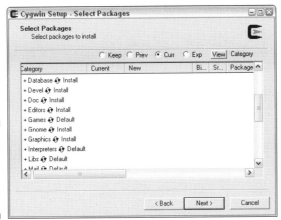

10. Select the packages you want to install. Click Next.

If you want them all, click Default beside the word All under Category.

The word *Default* appears next to many categories. Clicking this word more than once produces a range of results: Click it once, and it changes to *Install.* Click again, and it's *Reinstall.* Click it again, and it's *Uninstall.* Click one last time, and you are back at *Default.* We suggest you select Install. Installing everything takes up approximately 1 gigabyte. If you don't have the available space, select only those categories you think you will need.

If you choose to install everything, it can take a long time. Obviously, how long depends on the bandwidth of your connection to the Internet. It also depends on the speed of your processor. But trust us, when you install everything, it takes time, so prepare yourself for a long wait. Should you choose not to install a package at this time, you can always do so at a later time. Rerun the `setup.exe` program and install those programs you now want.

While Cygwin installs, the progress window shown in Figure 4-5 tracks your progress as it downloads the various components.

When the setup is complete, you see the window shown in Figure 4-6.

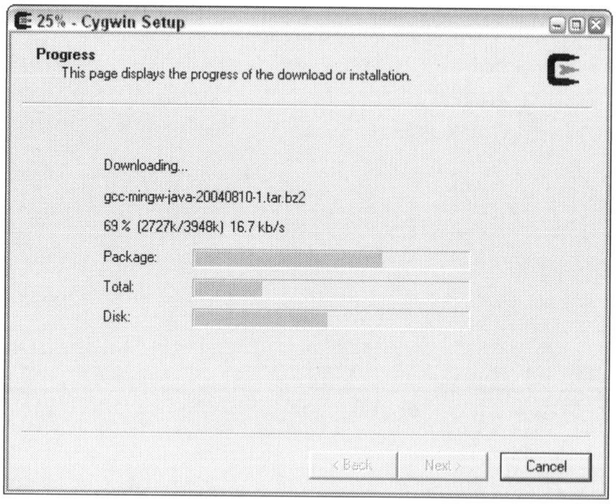

Figure 4-5:
Cygwin down-loading.

11. If you want to create desktop or Start menu icons, select (or deselect) the appropriate options. Click Finish.

That's it. You are now the proud owner of Cygwin.

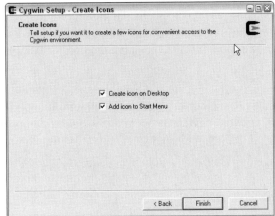

Figure 4-6:
Creating
desktop and
Start menu
icons.

Go to your desktop and double-click the Cygwin icon. Cygwin starts, and you
should see a window like the one shown in Figure 4-7.

Figure 4-7:
Cygwin
window.

Cygwin presents you with a command prompt. This is a bash shell. The Cygwin user is the same as the Windows user. If you want to see what Cygwin has mounted for you, in addition to the contents of the c:/cygwin directory you created, type df at the prompt. The c:/cygwin directory is the root directory.

You have the opportunity to try some of the UNIX tools in later chapters. But just to get started, type uname -a at the prompt. Try an ls -al command. Ever cursed Windows because you couldn't easily find out what processes are executing? Well, you just have to execute the ps -aWl command. (You might want to pipe (>) the output to a file.) If you're not familiar with UNIX commands, then you need to get a good UNIX book. Why not start with *UNIX For Dummies,* 5th Edition, by John Levine and Margaret Levine Young (Wiley)?

Cygwin has a couple of drawbacks:

- You have to use the UNIX version it gives you.
- You cannot run other operating systems.

That's a pretty short list considering that Cygwin is free (it is distributed under the GNU Public License). However, should you feel flush, you can move up to VMware.

Setting up VMware

VMware allows you to run simultaneous operating systems. The *VM* in VMware stands for *virtual machine.* You install a host operating system, such as Windows XP, and then install VMware Workstation on top of it. Then you install the guest operating system in VMware. The virtual machine is similar to your real machine: You can power it on and off, and it boots up just like the real thing. As a guest operating system, VMware allows you to install anything that runs on the Intel x86 architecture. This means you can install Solaris x86, Windows 2003 Server, Red Hat Linux, SUSE Linux, or any other operating system you choose. Still need to test Windows 98 programs? Use VMware. The only thing stopping you from running every operating system known to man is disk space and real memory.

You can download VMware from www.vmware.com. It takes up approximately 21MB.

Hover your cursor over the Products link at the top of the page and select the VMware Workstation link from the resulting drop-down list. If you click the red Buy Now button at the top, you go to the VMware Store, where you find out that VMware Workstation for Windows costs $189. After you use the software for a while, you'll agree this is a good price. (You can get a 30-day trial if you are not convinced.)

After you download VMware, it installs like any Windows application. Just follow the installation wizard.

During the download process, you might see a warning message to disable AutoRun. VMware doesn't like the CD-ROM AutoRun feature. (From a security standpoint, you shouldn't either.) Agreeing with VMware and disabling AutoRun is a good idea.

When the installation is complete, you need to reboot your machine. Now you are ready to add some guests or virtual machines. Installing new machines is easy:

1. **Start VMware.**

   You see a window like the one shown in Figure 4-8.

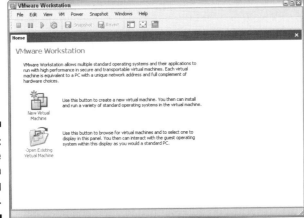

Figure 4-8:
VMware
Workstation
opening
window.

2. **Click the New Virtual Machine icon.**

   This starts the process of creating your first virtual machine. The New Virtual Machine wizard appears.

3. **Click Next.**

4. **Select Typical and click Next.**

   The Select a Guest Operating System window appears.

5. **Select the OS you want to install.**

   You have a choice of the following:

   • Microsoft Windows

   • Linux

   • Novell Netware

   • Sun Solaris

   • Other

If you select Other, you can install FreeBSD. Many good tools run on BSD.

If you select Linux, you can select a Linux version from the drop-down box.

6. **Select the version you have and click Next.**

7. **Type a name for your guest in the Virtual Machine Name box. Then click Next.**

   You can create any name you want, so pick one that is meaningful to you. Also, decide where you want to store the image. Leave the default unless you have a compelling reason not to do so.

8. **Select the Network Type. Click Next.**

   We suggest that you select Use Bridged Networking because it allows you to talk to your host operating system.

9. **Specify Disk Capacity.**

   Virtual machines have virtual disks. You can pick any size you want as long as you have the available space. We recommend you leave the default of 4GB and leave the two other boxes deselected.

10. **Click Finish.**

    However, you are not quite finished because you don't have a system image.

    You should see the window shown in Figure 4-9.

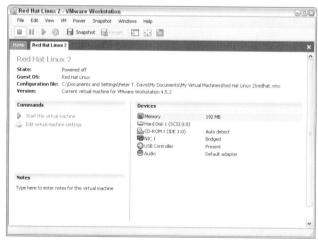

Figure 4-9:
Red Hat
Linux tab.

You now have a big choice. You can start the VM and install Red Hat Linux from a CD-ROM, or you can point to an ISO image. For this exercise, we'll do the latter.

11. **From the Commands panel, click Edit Virtual Machine Settings.**

    VMware presents the window shown in Figure 4-10.

12. **Click CD-ROM.**

    If you want to install the operating system from a CD, then skip to Step 14.

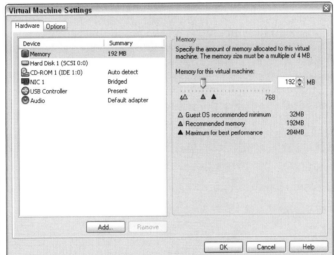

Figure 4-10:
Virtual
Machine
settings.

13. **From the right-hand pane, select Use ISO image.**

14. **Click the Browse button and find your ISO image. Click OK.**

15. **Click Start This Virtual Machine from the left-hand pane.**

    When you do this, you see a familiar display: The VM goes through the POST routine, does a memory check, and then boots itself.

Cygwin and VMware are wonderful tools, but you need to install them on your system; they won't run any other way. If you don't want to install software on your system, you can use products like Knoppix and WarLinux that boot from a diskette or a CD.

Linux distributions on CD

The following solutions are different from the partitioning and emulation solutions discussed above. What makes them different is that you don't need to install them on your system: They boot and run completely from a CD.

Knoppix, for instance, runs from a CD based on the Linux 2.6.*x* kernel. It is a free and Open Source GNU/Linux distribution. You don't need to install anything on a hard disk; it's not necessary. Knoppix has automatic hardware detection and support for many graphics cards, sound cards, SCSI and USB devices, and other peripherals. It includes recent Linux software, the K Desktop Environment (KDE), and programs such as OpenOffice, Abiword, The Gimp (GNU Image Manipulation Program), the Konqueror browser, the Mozilla browser, the Apache Web server, PHP, MySQL database, and many more quality open-source programs. Knoppix offers more than 900 installed software packages with over 2,000 executable user programs, utilities, and games.

You can download Knoppix (it is approximately 700 MB) or you can buy it from a CD distributor. Knoppix is available for download from `www.knoppix.net/get.php`. It's also included on a DVD in *Knoppix For Dummies* by Paul Sery (Wiley).

Knoppix is not the only distribution of Linux that fits on a CD. Consider also using one of the following Linux CD distributions:

- Cool Linux CD: `http://sourceforge.net/project/showfiles.php?group_id=55396&release_id=123430`
- DSL (Damn Small Linux): `www.damnsmalllinux.org`
- GNU/Debian Linux: `www.debian.org`
- SLAX: `http://slax.linux-live.org`
- WarLinux: `http://sourceforge.net/projects/warlinux`

  WarLinux is a special Linux distribution made for wardrivers. It is available on either a disk or bootable CD. The developer of WarLinux intended systems administrators to use it to audit and evaluate their wireless network installations.

Stumbling tools

In the methodology Kevin describes in his book, *Hacking For Dummies* (Wiley), and in the OSSTMM and ISSAF methods discussed in Chapter 2, the first step in ethical hacking is the same: reconnaissance. The best type of tool for reconnaissance is wardriving software. Programs like NetStumbler and Kismet help you find access points. Refer to Chapters 9 and 10 for more on the various stumbling tools.

You got the sniffers?

Stumbling tools help you find the access points, but that's not enough. You need to peek into the transmitted frames. If the frames are unencrypted, of course, then this is an easy task. But when the frames are encrypted, you

---

The origin of the word *sniffer*

The term *sniffer* came from a product called Sniffer that was manufactured and marketed by a company named Data General. Unfortunately for Data General, the name of their product has become the generic name for this type of software. (Ask the various companies whose products had the same thing happen to them — Aspirin, Kleenex, and Zipper — and they'll tell you this is not a good thing.) Still, you might hear these products referred to as network analyzers, data analyzers, protocol analyzers, packet analyzers, data line monitors, or network monitors.

---

need to decrypt the frame before you can look at it. This type of decryption software is generally called a *sniffer*.

Many freeware and commercial sniffer products are floating around out there. Some run on Windows, and others run on Linux. Two of the more popular sniffers are Ethereal and AiroPeek, which we cover in Chapter 8.

Picking Your Transceiver

Wireless Networks For Dummies (Wiley) provides information on the various form factors for your clients. You have lots of options to choose from. Picking your wireless network interface card or transceiver depends on the operating system you choose. When NetStumbler and Kismet first came out, there were two chipsets for wireless NICs: Hermes and Prism2. As a general rule, if you decide to use NetStumbler, you want a card based on the Hermes chipset. Kismet, on the other hand, works best with a Prism2 (Intersil) card. If you are prepared to do a kernel modification, then Hermes cards will work with Kismet.

Determining your chipset

Don't know whether you have a Prism2 chipset or a Hermes chipset? The following PC Card manufacturers use the Prism2 chipset:

- 3Com
- Addtron
- AiroNet
- Bromax
- Compaq WL100
- D-Link
- Farallon
- GemTek
- Intel
- LeArtery Solutions
- Linksys
- Netgear

- ✔ Nokia
- ✔ Nortel
- ✔ Samsung
- ✔ Senao
- ✔ Siemens

- ✔ SMC
- ✔ Symbol
- ✔ Z-Com
- ✔ Zoom Technologies

Further, if you have a Prism2 chipset, you may see a computer with antenna icon in the System Tray, as shown in Figure 4-11.

Prism2 icon

Figure 4-11:
Prism2 icon.

The following PC Card manufacturers use the Hermes (Lucent) chipset:

- ✔ 1stWave
- ✔ Agere/ORiNOCO/ Proxim
- ✔ Alvarion
- ✔ Apple
- ✔ ARtem
- ✔ Avaya
- ✔ Buffalo
- ✔ Cabletron

- ✔ Compaq WL110
- ✔ Dell
- ✔ ELSA
- ✔ Enterasys
- ✔ HP
- ✔ IBM
- ✔ SONY
- ✔ Toshiba

Much like the Prism2 chipset, if you have a Hermes (Lucent) chipset, you will see an icon in the System Tray, as shown in Figure 4-12.

Hermes icon

Figure 4-12:
Hermes
icon.

To find information for your Hermes chipset, visit `www.hpl.hp.com/personal/ Jean_Tourrilhes/Linux/Wireless.html` and look for "orinoco."

Buying a wireless NIC

When purchasing a wireless NIC, look for one that supports an external antenna. Figure 4-13 depicts an ORiNOCO card with an external antenna connector on the top. In this figure, the built-in antenna is the black plastic part on the end.

Figure 4-13: ORiNOCO Gold Classic card.

The ORiNOCO Gold Classic card from either Agere or Lucent is a popular card with wireless hackers because it has an external antenna connector and works with both Kismet and NetStumbler. Take care when buying new ORiNOCO cards. ORiNOCO is now owned by Proxim, which came out with an ORiNOCO card *not* based on the Hermes chipset. The Hermes card is still available, but it is usually sold as the ORiNOCO Gold Classic.

You can find a somewhat dated but useful comparison of the wireless cards and their chipsets at Seattle Wireless: `www.seattlewireless.net/index.cgi/HardwareComparison`.

Extending Your Range

Antennae are generally optional, but if you want to test the boundary of your wireless signal, they are a must. Many companies that sell PC wireless NIC cards also sell antennae. But many of these cards do not come equipped with a jack to plug in the antenna. Many people have resorted to modifying these PC cards to add jacks or soldering wires to the built-in antennas of their cards. Check out eBay for examples.

Other people are building antennas from everything from Pringles cans to PVC pipe. Peter has a tomato juice "cantenna" (see Figure 4-14). You can build your own or buy one starting at $39.95. Cantenna (www.cantenna.com) and Hugh Pepper (http://home.comcast.net/~hughpep) offer cantennae for sale. These are mainly directional designs, more commonly know as yagi-style antennas. They focus the 2.4 GHz wave, usually through a condenser (the can), to an element (a piece of copper) specifically placed in the antenna. This is why they are also called *wave guides.* The depicted antenna provides 16 dBi gain.

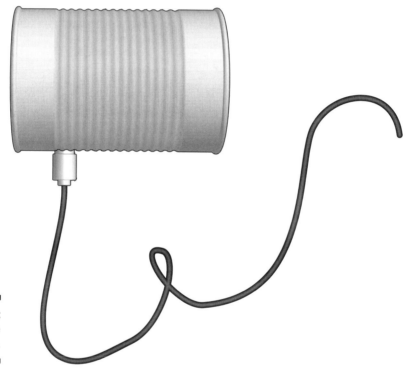

Figure 4-14: Wave guide cantenna.

Directional antennae are good for aiming at buildings across the street or pointing to the top of a very tall building, but they are not really good for wardriving. For wardriving, you want to get yourself an omnidirectional antenna like the one shown in Figure 4-15. Peter bought this 5 dBi antenna, which has a magnetic base that can be attached to a car or cart, on eBay for $5.95(!). At that price, you should buy several and give them as gifts.

TECHNICAL STUFF

Directional vs. omnidirectional antennas

We have actually had a great deal of success using directional antennas — as opposed to using omni antennas — for wardriving. If the directional antenna (or cantenna in this case) is aimed forward toward the front of the car, signals in front of you are often acquired much earlier than they are when using the omni antenna. The cantenna can then be moved left or right, peaking the signal and pointing out the exact origin or location of the wireless access point or errant signal being tracked.

With an omni, the signal strength gets stronger only as you get closer, but you can never be sure from which direction the signal is coming without actually traveling in several directions to track the signal strength. A directional antenna provides direction as well as signal strength when trying to locate a specific target. An omni can show a larger number of signals at one time than a directional antenna, but with lower signal strength than the directional antenna provides.

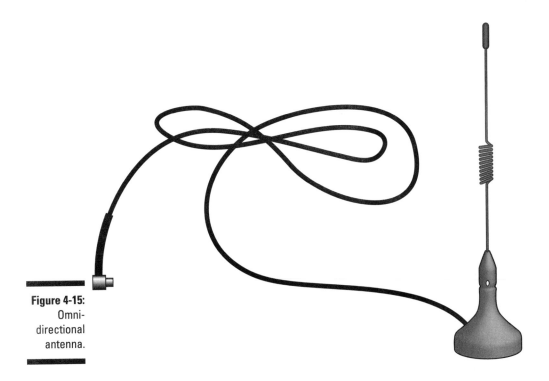

Figure 4-15: Omni-directional antenna.

For more information on antennae, we encourage you to check out *Wireless Networks For Dummies* (Wiley). That book outlines in depth the different types of antennae. You'll even find links for building your own wave guide antenna like that shown in Figure 4-16. That book provides information on RF mathematics so you can interpret what dBi means.

Using GPS

While driving in an unfamiliar place, Peter's family often asks, Where are we? Until he got his global positioning system (GPS), he couldn't always answer the question with great precision. As an answer, "somewhere between the Colorado border and El Paso" doesn't cut it, especially when you get close to restricted government areas. Now, with GPS, he can tell you the exact latitude and longitude. That GPS device can help with your wireless hacking efforts as well.

Using your GPS system with your wardriving software can give you more information. Remember, the hacker's primary law is *more information is better*. When you have to cover a large area in a short amount of time, the GPS is essential. Otherwise, you may not find the access point again.

To use GPS with wardriving software, you get the GPS unit to output GPS coordinates to the computer's serial port. When you find a wireless access point, Kismet and NetStumbler log the exact coordinates (down to a few feet) of the effective range.

Make sure you get a serial or USB cable to connect to your workstation when you buy your GPS device. If you are going to use the serial cable, ensure that you have a serial port; otherwise you'll need a serial-to-USB adapter. The standard protocol for GPS is NMEA (National Marine Electronics Association), which dumps your coordinates every 2 seconds to a serial port via a special cable at 9600,8,N,1. If you use a Garmin GPS, you can use the Garmin format. The Garmin eTrex Venture is nice for its size and cost (about $150). The Garmin reports every second, compared to every 2 seconds for the NMEA standard. However, Kismet supports only the NMEA format.

GPS units start at $100 and can run into the thousands. Peter purchased Microsoft Maps & Streets 2005 with GPS for about $129. The GPS (shown in Figure 4-16) labeled *Microsoft* is actually manufactured by Pharos, a well-known GPS vendor.

If you buy the Microsoft MapPoint software, you can take your output from NetStumbler and dump it right into StumbVerter (`www.michiganwireless.org/tools/Stumbverter`), which plots it on a map for you. You can then take your output, massage it, and import it into your Maps & Streets GPS device. We show you how to do this in Chapter 10. Table 4-1 lists some common mapping applications and their support for wardriving.

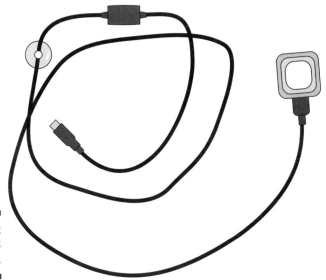

Figure 4-16:
USB GPS
device.

Table 4-1	**Mapping Software**		
Developer and Software	**GPS Interface Support?**	**Import "Pushpins"?**	**NetStumbler Support**
DeLorme Street Atlas USA	Yes	Yes	WiMap
DeLorme TopoUSA	Yes	Yes	PERL script
DeLorme XMap	Yes	Yes	PERL script
Microsoft MapPoint	Yes	Yes	StumbVerter
Microsoft Streets and Trips	Yes	Yes	PERL script

Signal Jamming

You can buy a transmitter and jam a signal, but jamming happens accidentally as well. Real crackers may jam your signals to deny service to your legitimate clients. The following can affect your signal:

A brief history of jamming

Deliberate radio jamming goes back to the beginning of the last century. A betting person would probably wager that radio jamming was first done by the military somewhere. That betting person would lose. The first recorded instance of jamming occurred during the America's Cup yacht races. Someone was trying to get a competitive leg up, not a military edge.

✔ Cordless phones can cause narrowband interference, which may mean you need to eliminate the source.

✔ Bluetooth devices and microwave ovens can cause all-band interference, which may mean you need to change the technology or eliminate the source.

✔ Lightning can charge the air, which may mean you need to ground and protect your equipment.

So, random interference can result in denial of service, but someone can do it intentionally as well by using one of two types of RF jamming devices:

✔ **RF generators** are rather expensive devices. You can get RF generators from companies like HP (www.hp.com) and Anritsu (www.anritsu.com).

✔ **Power signal generators** (PSGs) are not as pricey. They are used to test antennae, cables, and connectors. You can get PSGs from YDI (www.ydi.com) and Tektronix (www.tek.com).

A variety of jammers are complete, standalone systems consisting of appropriate antennae, energy sources, and modulation electronics, such as techniques generators. But what causes interference is the effective radiated power (ERP). You can use the jammer to disrupt the operation of electromagnetic systems in either receiving or transmitting modes to reduce or deny the use of portions of the electromagnetic spectrum.

You may want to test your wireless network to discover how susceptible it is to signal jamming from outside your organization. If you plan to run mission-critical applications over wireless networks, then you need to know whether others can cause unplanned network outages.

Part II
Getting Rolling with Common Wi-Fi Hacks

The 5th Wave   By Rich Tennant

"We need to pimp our security system."

In this part . . .

It's time to roll! With Part I under your belt, you're ready to move into the testing phase of your wireless-hacking efforts. This part starts out with a discussion of human insecurities — that is, things your users do that make your wireless networks vulnerable. Then we get into physical security vulnerabilities, common wireless client weaknesses, and default settings that can leave your systems begging to be attacked. We finish off with an introduction to wardriving — that's where the true fun begins. Stick around. It only gets better from here.

Chapter 5

Human (In)Security

*W*hen people think about the various vulnerabilities of wireless networks, they often concentrate only on the technical vulnerabilities associated with things like WEP, radio signal leakage, and the potential for DoS (denial of service) attacks. However, higher-level wireless-network vulnerabilities are often overlooked — namely, human insecurities. We're not talking about people with low self-esteem, but insecurities involving users who are ignorant of wireless risks or careless with wireless networks.

What kinds of problems can this cause? For starters, employees, and even trusted outsiders such as contractors and auditors, can bring in their wireless APs and plug them right into your network. Some may even set up their wireless clients to run in peer-to-peer — or *ad hoc* — mode with each other, which can pose even more risks to your network.

Adding to the mess, wireless systems are often implemented with the default settings by unknowing users. Hackers love this. Most of these users aren't malicious, they're just uninformed. But setting up unapproved wireless devices without considering the first bit of security is the last thing you need. And users, or even wireless network administrators, often unknowingly choose weak passwords, making systems even more vulnerable.

Another good way for hackers to break in to your systems is through the use of *social engineering*. This is when someone poses as a legitimate person (employee, consultant, or government official) and exploits the trusting nature of humans for ill-gotten gains.

Human fallibility is arguably the greatest threat to your systems, and these types of non-technical weaknesses are often the root cause of most

Melts in your mouth, not in your hands

Some networks are said to have *candy security*. This occurs when the wireless network has a hard, crunchy outside for protection (WEP, secure authentication, and so on) but falls short on security due to its soft, chewy inside (your people and weak processes). Make sure you avoid this problem.

wireless-network risks. This chapter explores these weaknesses and shows you how best to eliminate them.

What Can Happen

New wireless vulnerabilities come and go, and securing against unknown threats and vulnerabilities is very difficult. However, one thing's for sure: When the human element is introduced into information systems (and when is it not?), vulnerabilities start popping up everywhere and often remain indefinitely.

 The big picture must not be forgotten. In fact, securing the technical piece is pretty easy — it's securing the human element that takes more time and effort. Remember that both types of security must be accounted for. Otherwise, you're running a partially secured wireless network that can provide only limited information security.

What sorts of things can happen when human vulnerabilities are ignored? Well, for starters, things like this:

✔ Managers and network administrators deploy wireless network connectivity just because it's the latest and greatest technology or to appease their users who think it'd be neat to have all without considering the security issues or consequences involved with their actions.

✔ Social engineers work their way into your building or computer room.

✔ Users install APs for the sake of convenience and end up bypassing security controls, extending your network, and letting in unauthorized users without your knowledge.

✔ Hackers or malicious insiders exploit physical security weaknesses, leading to theft, reconfiguration of APs, cracking of WEP keys, and more.

✔ Network administrators and security managers deploy wireless networks with security requirements that are too stringent, which leads to users ignoring policies and bypassing controls any chance they get.

The possibilities are limitless.

Ignoring the Issues

We're heading toward a wireless world in which we'll have as much wireless traffic as wired traffic, if not more. The demand for "anywhere all-the-time" wireless network access, from the boardroom to the coffee shop, is continually growing. The bad thing is that many wireless networks are being deployed without concern for the big picture. The long-term consequences of insecurely implementing wireless systems are being ignored from the get-go.

One of the best things IT professionals can do is to consider security at the ground level before installing any type of system. If wireless networks are put in place with security in mind, it's much easier to make security changes long-term.

Most users, many business executives, and even some administrators ignore warnings that 802.11-based wireless networks are inherently insecure. By now, anyone watching television, reading the paper, or even reading their wireless network user's guide should know that simply connecting a wireless AP to the network without enabling any of the basic security features can have a negative impact on information privacy and security. However, as we often see, the desire for unlimited wireless connectivity usually outweighs any potential risks.

In the ongoing battle of security versus convenience and usability, what's secure is often not convenient or very usable for the user, and what's convenient, feature-rich, and user-friendly is often not secure. This mindset is what leads to many wireless network exploits.

Hotspots are now all the rage. Everyone wants connectivity and ease of use, and security is often pushed aside. What most users don't realize is just how insecure their computers and data are when they connect to an unsecured wireless network. Many people just connect to whatever AP is available, especially if they're out of the office, without thinking about the consequences. Making matters worse, newer, more "user-friendly" operating systems such as Windows XP make wireless network connection even more dangerous because the computer automatically connects to the first wireless network it sees — yours, theirs, or someone else's.

Common excuses for setting up unauthorized wireless networks are "I didn't know wireless security was such a problem," or "management just won't buy into the costs associated with securing our wireless network." However, the constant deluge of new information exposes the truth: 802.11-based systems *can* be made very secure with minimal money, time, and effort.

Not all users make the wireless security mistakes we speak of. However, the general tendency is to get things up and running as quickly as possible, overlooking what really needs to be done to secure 802.11-based networks.

Still, study after study shows that a large portion — quite often the majority — of wireless networks don't even utilize the most basic security features, such as WEP encryption and SSIDs, other than the defaults. Our work on ethical-hacking projects confirms these findings.

The only way to fix this problem is to change the mindset of general computer users, and that means educating users about security vulnerabilities that they might not even realize. Let's jump right in and look at some specific non-technical vulnerabilities you can test for.

Social Engineering

Social engineering is a technique used by attackers to take advantage of the natural trusting nature of most human beings. Criminals often pose as an insider or other trusted person to gain information they otherwise wouldn't be able to access. Hackers then use the information gained to further penetrate the wireless and quite possibly the wired network and do whatever they please.

Social engineering shouldn't be taken lightly. It can allow confidential or sensitive information to be leaked and cause irreparable harm to jobs and reputations. Proceed with caution and think before you act.

Social engineering is more common and easier to carry out in larger organizations, but it can happen to anyone. Testing for social-engineering exploits usually requires assuming the role of a social engineer and seeking vulnerabilities by approaching people and subtly probing them for information. If your organization is large enough that most people won't readily recognize you, carrying out the tests yourself should be pretty easy. You can claim to be a

- ✔ Customer
- ✔ Business partner
- ✔ Outside consultant or auditor
- ✔ Service technician
- ✔ Student at a university

If there's any chance of being noticed, or if you simply don't feel comfortable doing this type of testing, you can always hire a third party to perform the tests we talk about in this section. Just make sure you hire a trusted third party, preferably someone you've worked with before. Be sure to check references, perform criminal background checks, and have the testing approved by management up front.

The key is to look at this from a hacker's perspective. Outside of the technical methods we describe elsewhere in this book, ask yourself how a malicious outsider could gain access to your wireless network. The options and techniques are limitless.

Passive tests

The easiest way to start gathering information you can use during your social engineering tests is to simply search the Internet. You can use your favorite search engine to look up public information such as phone lists, organizational charts, network diagrams, and more. You can then see, from an outsider's perspective, what public information is available that can be used as an inroad for social engineering and ultimate penetration into your network.

One of the best tools for performing this initial reconnaissance is Google. It's amazing what you can do and find with Google. It's even more amazing that this information is made accessible to the public in the first place! You can perform generic Google queries for keywords and files that could lead to more information about your organization and network. Be sure to do both a Web and Groups search in Google because they may both contain some interesting information.

You can also perform some more advanced Google queries that are specific to your network and hosts. Simply enter the following directly into Google's search field to look for information that could be used against you:

- ✔ `site: your~public~host~name/IP` *keywords to search for*

  Look for keywords such as *wireless, address, SSID, password, .xls* (Excel spreadsheets), *.doc* (Word documents), *.ppt* (Power Point slides), *.ns1* (Network Stumbler files), *.vsd* (Visio drawings), *.pkt* (sniffer packet captures), and so on.

- ✔ `site: your~public~host~name/IP filetype:ns1 ns1`

  This searches for Network Stumbler files that contain wireless network configuration information. You can perform this query on any type of file, such as .vsd, .doc, and so on.

- ✔ `site: your~public~host~name/IP inurl:"h_wireless_11g.html"` or `inurl:"ShowEvents.shm"`

  This searches publicly accessible APs (yikes!) such as D-Link and Cisco Aironet for wireless setup pages and event logs, respectively. You may not think your systems have such a vulnerability, but do this test — you may be surprised.

These are just a few potential Google queries you can perform manually, just to get you started. Be sure to perform these queries against all of your publicly accessible hosts. If you're not sure which of your servers are publicly accessible, you can perform a ping sweep or port scan from outside your firewall to see which systems respond. (This is not foolproof because some systems don't respond to these queries, but it's a good place to start.)

For in-depth details on using Google as an ethical-hacking tool, check out Johnny Long's Web site, `http://johnny.ihackstuff.com`. This site has a wealth of information on using Google for advanced queries. It also includes a query database, called the Google Hacking Database (GHDB), where you can run various queries directly from the site.

You can also run automated Google tests in-house using a neat tool by Foundstone called SiteDigger. This tool, which is available at `www.foundstone.com/resources/freetools.htm`, allows you to run various pre-packaged Google queries against your systems — including the ones from Johnny Long's GHDB — as well as custom queries you make up yourself. The only limitation to this is that the Google API license required to run these tests permits a maximum of 1,000 Google queries per day. This limitation, however, is often more than you need. Figure 5-1 shows the user interface for SiteDigger version 2.0.

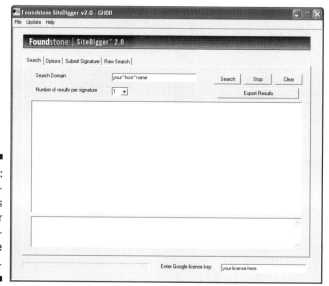

Figure 5-1: Foundstone's SiteDigger for automating Google queries.

Active tests

You can use various methods to go about gathering information from insiders. Two simple and less in-your-face methods are e-mail and the telephone. Simply pick up the phone, make a call to the help desk or to a random user, and start asking questions. Use a phone on which your caller ID won't give away your identity, such as a phone in the reception area or break room, a pay phone, or perhaps a colleague's office. You can even use your own phone if you think your users are gullible enough or won't recognize your name or number. You can do the same with e-mail. Change your e-mail address in your e-mail client (if possible) or use an obscure Webmail account and pose as an outsider.

A common method of social engineering is to gain direct physical access to wireless clients and APs. However, the good thing (or bad thing, depending on how you look at it) about wireless networks is that physical access is not necessary. Chapter 6 covers the physical aspects of wireless security in depth.

You can also just show up in person, acting as an outsider. Whichever method you choose, your goal is to glean information from employees and other users on your network that would essentially give you the "keys" you need for gaining external access to the wireless network. This includes:

- SSIDs
- WEP key(s)
- Computer and network login passwords
- Preshared secret passphrases used by authentication systems such as WPA
- Legitimate MAC or IP addresses used to get onto the network

You could call up your help desk or any random user, pose as a legitimate employee or business partner, and ask for wireless configuration information such as the SSID or WEP key(s). You can ask practically anyone for this information. They may

- Know it off the top of their head
- Have it written down and readily available
- Let you walk them through looking the information up on their computer
- Refer to someone else who can help

After you gather as much information as you feel comfortable gathering, you should check to see just how far you can penetrate the network as an outsider.

Unauthorized Equipment

A very common problem network administrators and security managers face is the introduction of unauthorized wireless systems onto the network. Some users — especially those who are technically savvy — don't like to be told they can't use wireless network technology in their workspace, so they may take the initiative to do it themselves, often in direct defiance of organizational policy.

You can even have a malicious insider or, worse, an outsider on an adjacent floor, who has set up a rogue AP for your users to connect to. This is a very simple setup for the hacker. All he has to do is set up an AP using your SSID and wait for your wireless systems to associate with it. There are also programs that automate the process of creating "fake" APs. If this occurs, hackers can capture virtually all traffic flowing to and from your wireless clients. We cover this in more depth in Chapter 11.

A more common problem is the naïve introduction of wireless systems by users who either don't understand the security issues associated with their actions or aren't aware of company policies. Either way, you've got a potential mess on your hands.

Let's take a look at an unauthorized AP scenario. When it comes to users installing unauthorized wireless systems, here's how it usually happens:

1. An employee, Lars, wants to be able to work on his laptop in an adjacent, more plush, cubicle. However, that cubicle doesn't have an Ethernet network drop.

2. Lars thinks of a solution: 'Instead of dealing with IT to get a new drop installed or asking them to come up with another solution, I can just install a wireless AP in my main work area and communicate wirelessly from my laptop to the network!'

3. Lars strolls merrily down to the local consumer electronics store during his lunch break and buys a "wireless-network-in-a-box" solution. What a deal — he can get an AP, a wireless PC Card for his laptop, and 5,000 free hours on AOL for the low price of $59.95. Subtracting the $50 in mail-in rebates, Lars has a newfound freedom from network cabling for only $9.95!

4. Lars returns to the office, unpacks his treasure, plugs the AP into the network jack in his original cubicle, and installs the wireless NIC in his laptop.

5. Lars powers up the AP, which, in typical fashion, has a valid IP address for your network preprogrammed into it. Remember, to make things convenient for the end users, no security settings are enabled on the AP — no WEP, broadcasting of the default SSID, blank admin password — nothing. He thinks to himself, 'Wow, who would've thought it'd be this easy!?'

6. Lars boots his laptop, which grabs an IP address from the AP that is running its own DHCP server, and he's off! He's now able to log on to your network and browse the Internet. Again, Lars can't believe how easy this was to set up and thinks that maybe IT is his calling.

Total elapsed time: 45 minutes. Consequences of Lars's actions: *Complete* and *unlimited* exposure of your network to the outside world.

This is a typical scenario, and it didn't require a whole lot of know-how on Lars's part. But some people are savvier. They know that they don't need an AP to communicate with other wireless users directly. These peer-to-peer or ad hoc systems can be even trickier to track down because no AP is involved.

We often hear "my users wouldn't do that" or "I know my network," but believe it or not, regardless of the size of the organization, this scenario happens very easily and very often.

If you're on a limited budget and want to get a general view of wireless APs in your building, you can use a wireless laptop running Windows XP. Here's a quick test you can run to look for unauthorized APs and wireless clients before they get the best of your network:

1. **On the Windows XP desktop, right-click My Network Places and select Properties.**

   The Network Connections window opens.

2. **Double-click your wireless network card.**

   The Status window opens.

3. **Select View Wireless Networks.**

You can walk around your building to see what comes up. Unfortunately, in order for new APs to show up, you have to click Refresh Network List in the upper-left corner of the window, or simply press F5 on your keyboard.

Figure 5-2 shows an example of what this looks like. Notice how one AP shows up with the Lock icon labeled *Security-enabled wireless network,* and the other two (including Lars') don't. The one that has security enabled is using WEP encryption. The other two (including Lars') are, well, wide open. Shame on Lars!

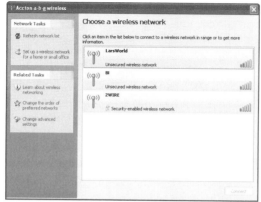

Figure 5-2:
Browsing
for available
wireless
networks in
Windows
XP.

In the name of privacy and protecting the innocent, in Figure 5-2 and many other figures throughout the book, we cropped MAC addresses and other wireless information from the screenshots.

For this kind of testing, you can also use the software that comes with your wireless NIC. These programs often offer greater details about the wireless systems found. For instance, ORiNOCO's Client Manager has a feature called Site Monitor that allows you to browse your airwaves and view such settings as MAC addresses; signal-to-noise ratios (SNR), which can indicate how close you are to the wireless device; and specific radio channels being used. Added bonuses include a logging feature and the fact that you don't have to continuously hit refresh for updated information, as you do with the generic Windows XP management software.

One caveat to all this is that many APs can be configured so that their SSIDs are not broadcast and 802.11 beacon packets — packets APs use to advertise their presence — are sent out only every minute or so. This helps hackers keep their rogue systems from showing up on client management and stumbling software. Because the main focus is on the average user setting up an AP, this is not really an issue to worry about here. We cover more advanced rogue AP detection in Chapter 11.

Default Settings

Although we dedicate an entire chapter to the topic of default wireless settings (Chapter 8), they deserve mention here because of the human issues surrounding them. An unbelievable number of APs are deployed with the default settings still intact, including, for example:

- IP addresses
- SSIDs

✔ Broadcasting of SSIDs

✔ Admin passwords

✔ Remote management enabled

✔ Full power settings

✔ Use of omnidirectional antennas that come standard on most APs

✔ No MAC-address filtering

✔ WEP turned off

There are also related updates to AP firmware as well as client management software and drivers that come with the wireless systems. Wireless vendors are continually updating their firmware and software to fix security vulnerabilities and add enhanced security features, yet patching and updating is often overlooked.

Hackers know they can download the documentation for practically any 802.11-based wireless network right off the Internet. This documentation often reveals many of the default settings in use. In addition, several independent Internet sites list default settings, including:

✔ www.cirt.net/cgi-bin/passwd.pl

✔ www.phenoelit.de/dpl/dpl.html

✔ http://new.remote-exploit.org/index.php/Wlan_defaults

✔ www.thetechfirm.com/wireless/ssids.htm

If you want to see if your users or any of the systems you've set up are using vulnerable default settings, you can perform some basic tests with the information you've gathered, including

✔ Connecting to APs by using their default SSIDs

✔ Remotely connecting to the default admin port

✔ Spoofing MAC addresses (we cover this in detail in Chapter 13)

Refer to Chapter 8 for details of the various default setting tests you can perform against your network.

Weak Passwords

The use of weak passwords on wireless systems is a major problem. Passwords are often one of the weakest links in the information-security chain — especially on wireless networks, where they're easier to glean and crack. From remote admin access to WEP to WPA preshared secrets to wireless client operating systems, passwords can be the Achilles heel of your network in quite a number of ways.

It's easy to create and maintain strong passwords that are very difficult to crack, although users often neglect this. A single weak password can cause a big problem. If a hacker gains access to a password on the wireless network, all bets are off, and bad things usually start happening.

An effective password is one that's both difficult to guess yet easy to remember.

The highly publicized encryption flaws inherent in the WEP protocol have also been an impediment to more widespread use of wireless networks. A not-so-determined hacker only has to capture a day's worth of wireless packets — often less — in order to use various cracking tools to determine your WEP key. As we mentioned before, WPA and WPA2 have solved all the known WEP issues. But they have their own problems as well! And most wireless networks are not using WEP, so hackers are still breaking in. WEP is not completely worthless, though, because it still provides a layer of security — a hoop if you will — that an attacker has to jump through to get to your systems.

We cover the topic of weak passwords in other chapters throughout the book, including Chapter 7 on wireless clients, Chapter 15 on WEP, and Chapter 16 on authentication. Kevin also discusses passwords in depth in his passwords chapter in *Hacking For Dummies*. If you haven't yet purchased *Hacking For Dummies* but you're just dying to learn more about password hacking, you can download the password chapter for free at `http://searchsecurity.tech` `target.com/searchSecurity/downloads/HackingforDummiesCh07.pdf`.

Human (In)Security Countermeasures

You can combat the human insecurities your wireless network faces in several ways. These come in the form of policy, education, proactive monitoring, and simple prevention. The solutions are fairly straightforward. The real trick is getting users, and most importantly, upper management to buy into them. Here's what you can do.

Enforce a wireless security policy

The first step is to create a company policy that no unauthorized wireless systems are to be installed. The following is an example of a wireless policy statement:

Users shall not install or operate any wireless-network system (router, AP, ad-hoc client, etc.) within the organization.

If you choose to allow wireless systems inside your organization or allow remote users to have wireless networks at home, your wireless security policy should outline specific minimum requirements. The following is an example of such a policy:

Users shall not install or operate any wireless-network system (router, AP, ad-hoc client, etc.) within the organization without written permission from the Information Technology Manager. Additionally, all wireless systems must meet the following minimum requirements:

- ✔ *WEP is enabled.*
- ✔ *Default SSIDs are changed to something obscure that doesn't describe who owns it or what it is used for.*
- ✔ *Broadcasting of SSIDs is disabled.*
- ✔ *Default admin passwords are changed to meet the requirements of organizational password policy.*
- ✔ *APs are placed outside the corporate firewall or in a protected DMZ.*
- ✔ *Personal firewall software such as Windows Firewall or BlackICE is installed and enabled.*

Train and educate

One of the best ways to get users to adhere to your wireless security policy is to make them aware of it — teach them what the policy means, along with the consequences of violating the policy. Educate users on what can happen when the policy is not adhered to and try to relate these issues to their everyday job tasks. For example, where a project manager is using a wireless network, describe to her how a hacker could capture detailed information about the project she's working on, such as user lists, network diagrams, costs, and other confidential information.

If management doesn't get user sign-off on your policies showing that they understand and agree to the terms of the policies, the policies are as good as nothing. Make sure sign-off takes place.

Also, talk to your users about how a hacker can make it look like the user actually committed the crime by spoofing the user's address, using the user's login information, sending e-mails on the user's behalf, and so on.

Keep people in the know

If you want to keep security on top of everyone's minds, the training and awareness has to be ongoing. Keep people aware of security issues by passing out items (such as the following) with security messages on them:

- ✔ Screen savers
- ✔ Mouse pads
- ✔ Pens and pencils
- ✔ Sticky-note pads
- ✔ Posters in the break room

Several organizations specialize in these security awareness products. Check out

- ✔ www.securityawareness.com
- ✔ www.thesecurityawarenesscompany.com
- ✔ www.greenidea.com
- ✔ www.privacyposters.com

Your best defense is your people, so keep them in the know and make sure you put a positive spin on your security initiatives so you don't tire them out.

Scan for unauthorized equipment

A great way to help enforce your wireless security policy is to install a centrally managed wireless gateway or IDS system, such as the products offered from Bluesocket (www.bluesocket.com) and AirDefense (www.airdefense.net). These systems can prevent problems from the get-go through strong authentication or alerts when they detect unauthorized wireless systems, can monitor for malicious wireless behavior, and more. We outline how to get similar functionality out of other tools such as commercial monitoring programs and wireless sniffers in Chapter 11.

Secure your systems from the start

Another great defense against people-related security vulnerabilities on your wireless network is to prevent them in the first place. Set your users and your systems up for success. You should not only make it policy to harden wireless systems but also help users do the hands-on work if possible. Also, ongoing ethical hacks and audits (comparing what is *supposed* to be done according to policy to what is *actually* being done) are essential. This can help you make sure that wireless systems haven't been changed back to include the insecure settings you're trying so hard to prevent.

Chapter 6

Containing the Airwaves

In This Chapter

▶ Monitoring link strength and quality

▶ Choosing monitoring software

▶ Protecting your organization

Many companies expose themselves to attack because they don't attempt to control the radio signals leaking from their organization. In such cases, a cracker could sit in your parking lot or stand across the street and monitor your network. This chapter shows you how to control your signals. In later chapters, we show you how to monitor frames (Chapter 8), discover networks (Chapters 9 and 10), intercept frames (Chapter 12), deny service (Chapter 13), crack encryption keys (Chapter 14), and beat user authentication (Chapter 15). Before you can try these tests, you need to find radio signals — yours and others.

Signal Strength

A first step to testing your network is to determine the bounds of your network. You can use sophisticated tools like AiroPeek (see Chapter 8) or a spectrum analyzer, but that would really be overkill. All you need are various software programs that supply link-quality information. Several freeware products run on Linux.

Using Linux Wireless Extension and Wireless Tools

The Linux Wireless Extension and Wireless Tools are an open source project sponsored by Hewlett Packard. The Wireless Extension is a generic application programming interface (API) that gives you information and statistics about the user space. Wireless Tools is a set of tools that use the Wireless Extensions. The Wireless Tools are

- ✔ **iwconfig:** Changes the basic wireless parameters.
- ✔ **iwpriv:** Changes the Wireless Extensions specific to a driver (private).
- ✔ **iwlist:** Lists addresses, frequencies, and bit rates.
- ✔ **iwspy:** Gets per-node link quality.

We explore these tools in turn in the following sections. For each tool, we provide an illustrative example. If you want to really understand the command and its many parameters, however, please check out the man page for the syntax and other information about any of these commands. If you have a Web browser, you can use Google.

Linux Wireless Extensions are powerful additions to your ethical hacking utility belt. Linux Wireless Extensions are available from `http://pcmcia-cs.sourceforge.net/ftp/contrib`. Look for the entry wireless_tools.27.tar.gz near the bottom of the available documents and programs. Wireless Extensions v.14 is bundled in the 2.4.20 kernel, and v.16 is in the 2.4.21 kernel.

`iwlist` and the others are great tools. They get their information from the standard kernel interface `/proc/net/wireless`. But these tools provide only a snapshot in time; they do not provide statistics over time. If you favor the Windows platform, you can use a great tool like NetStumbler (we cover this tool in depth in Chapter 9). But when you work with Linux, you want to find a better link-monitoring tool. The other tools in this section provide more functionality than `iwconfig`, `iwpriv`, `iwlist`, and `iwspy`.

Using iwconfig

You use `iwconfig` to configure a wireless network interface. If you're familiar with the `ifconfig` command, the `iwconfig` command is similar but works only with wireless interfaces. You use `iwconfig` to set the network interface parameters, such as frequency. As well, you can use `iwconfig` to set the wireless parameters and display statistics. The syntax is as follows:

```
iwconfig interface [essid X] [nwid N] [freq F] [channel C]
              [sens S] [mode M] [ap A] [nick NN]
              [rate R] [rts RT] [frag FT] [txpower T]
              [enc E] [key K] [power P] [retry R]
              [commit]
iwconfig --help
iwconfig --version
```

Let's look at each one of the parameters.

- ✔ **essid:** Use the ESSID parameter to specify the ESSID or Network Name. For example, the following specifies that you want to set the ESSID for the wireless adapter to ANY for wardriving.

```
iwconfig eth0 essid any
```

✔ **nwid/domain:** Use the Network ID parameter to specify the network ID or Domain ID. For example, the following specifies that you want to disable Network ID checking.

```
iwconfig eth0 nwid off
```

✔ **freq/channel:** Use this parameter to set the operating frequency or channel. A value below 1,000 represents the channel number, while a value over is the frequency in Hz. For example, the following specifies that you want to set the frequency to 2.422 GHz.

```
iwconfig eth0 freq 2.422G
```

Or for example, the following specifies that you want to use channel three.

```
iwconfig eth0 channel 3
```

✔ **sens:** Use this parameter to set the sensitivity threshold. For example, the following specifies the level as 80 dBm.

```
iwconfig eth0 sens -80
```

✔ **mode:** Use this parameter to set the operating mode of the device. The operating mode is one of the following:

- **Ad-hoc:** no Access Point.
- **Managed:** more than one Access Point, with roaming.
- **Master:** synchronization master or an Access Point.
- **Repeater:** node forwards packets between other wireless nodes.
- **Secondary:** node acts as a backup master or repeater.
- **Monitor:** the node acts as a passive monitor and only receives packets.
- **Auto:** self-explanatory.

For example, the following specifies that the network is infrastructure mode.

```
iwconfig eth0 mode managed
```

✔ **ap:** Use this parameter to force the card to register to the Access Point given by the address. Use off to re-enable automatic mode without changing the current Access Point, or use any or auto to force the card to re-associate with the current best Access Point. For example, the following forces association with the access point with the hardware address of 00:60:1D:01:23:45.

```
iwconfig eth0 ap 00:60:1D:01:23:45
```

✔ **nick[name]:** Use this parameter to set the nickname or station name. For example, the following sets the nickname to Peter Node.

```
iwconfig eth0 nickname Peter Node
```

✔ **rate/bit[rate]:** Use this parameter to set the bit-rate in bits per second for cards supporting multiple bit rates. For example, the following sets the bit rate to 11 Mbps.

```
iwconfig eth0 rate 11M
```

✔ **rts[_threshold]:** Use this parameter to turn RTS/CTS on or off. For example, the following turns RTS/CTS off.

```
iwconfig eth0 rts off
```

✔ **frag[mentation_threshold]:** Use this parameter to turn fragmentation on or off. For example, the following specifies a maximum fragment size of 512K.

```
iwconfig eth0 frag 512
```

✔ **key/enc[ryption]:** Use this parameter to turn encryption or scrambling keys on or off and to set the encryption mode. For example, the following specifies an encryption key.

```
iwconfig eth0 key 0123-4567-89
```

✔ **power:** Use this parameter to set the power management scheme and mode. For example, the following disables power management.

```
iwconfig eth0 power off
```

✔ **txpower:** Use this parameter to set the transmit power in dBm for cards supporting multiple transmit powers. For example, the following set the transmit power to 15 dBm.

```
iwconfig eth0 txpower 15
```

If you are unfamiliar with dBM as a measurement, refer to www.atis.org/ tg2k/_dbm.html for a definition.

✔ **retry:** Use this parameter to set the maximum number of MAC retransmission retries. For example, the following specifies to retry 16 times.

```
iwconfig eth0 retry 16
```

✔ **commit:** Use this parameter to force the card to apply all pending changes rather than waiting for the issuance of an ifconfig command. For example, the following specifies to commit the changes.

```
iwconfig eth0 commit
```

✔ **Link quality:** Use this parameter to display the quality of the link.

✔ **Signal level:** Use this parameter to show the received signal strength.

✔ **Noise level:** Use this parameter to display the background noise level.

✔ **invalid nwid:** Use this parameter to detect configuration problems or the existence of an adjacent network.

✔ **invalid crypt:** Use this parameter to display the number of packets that the hardware couldn't decrypt.

✔ **invalid misc:** Use this parameter to display other packets lost in relation with specific wireless operations.

There you have it. Remember you can get more information by using the man command.

Using iwpriv

iwpriv is the companion tool to iwconfig. Again, you use iwpriv to configure optional (private) parameters for a wireless network interface. You use iwpriv for parameters and settings specific to each driver, as opposed to iwconfig, which deals with generic ones. The syntax is as follows:

```
iwpriv interface private-command [I] [private-parameters]
iwpriv interface -all
iwpriv interface roam {on,off}
iwpriv interface port {ad-hoc,managed,N}
```

Using the iwpriv command without any parameters lists the available private commands for each interface and the parameters required.

Let's look at each one of the parameters.

✔ **private-command [I] [private-parameters]:** Use the specified private-command on the interface. The I parameter, which stands for an integer, is the integer to pass to the command as a Token Index. Your driver documentation should specify the value for the integer, otherwise leave the value out.

The command may optionally take or require arguments, and may display information. The following paragraphs provide information on the arguments.

✔ **-a/--all:** Use this parameter to execute and display all the private commands that don't require any arguments, for example, read only.

✔ **roam:** Use this parameter to enable or disable roaming, when supported.

✔ **port:** Use this parameter to read or configure the port type.

Using iwlist

`iwlist` allows you to display more detailed information from a wireless interface than you can get with `iwconfig`. For instance, you can get the ESSID, node name, frequency, signal quality and strength and bit data and error rate. The syntax is as follows:

```
iwlist interface scanning
iwlist interface frequency
iwlist interface rate
iwlist interface key
iwlist interface power
iwlist interface txpower
iwlist interface retry
iwlist --help
iwlist -version
```

Let's look at each one of the parameters.

- ✔ **scan[ning]:** Use this parameter to specify the access points and ad-hoc cells in range. For example, the following enables scanning.

  ```
  iwlist wlan0 scan
  ```

  Run this command and you may see something like the following:

  ```
  wlan0    Scan completed:
        Cell 01 - Address: 00:02:2D:8F:09:8D
              ESSID:"pdaconsulting"
              Mode:Master
              Frequency:2.462GHz
              Quality:0/88 Signal level:-50 dBm Noise level:-
        092 dBm
              Encryption key:off
              Bit Rate:1Mb/s
              Bit Rate:2Mb/s
              Bit Rate:5.5Mb/s
              Bit Rate:11Mb/s
  ```

- ✔ **freq[uency]/channel:** Use this parameter to specify the list of available frequencies for the device and the number of defined channels.

- ✔ **rate/bit[rate]:** Use this parameter to list the bit-rates supported by the device.

- ✔ **key/enc[ryption]:** Use this parameter to list the supported encryption key sizes and to display all the available encryption keys.

- ✔ **power:** Use this parameter to list the various Power Management attributes and modes of the device.

- ✔ **txpower:** Use this parameter to list the various Transmit Powers available on the device.

- ✔ **retry:** Use this parameter to list the transmit retry limits and retry life-time on the device.

- ✔ **--version:** Use this parameter to display the version of the tools, as well as the recommended and current Wireless Extensions version for the tool and the various wireless interfaces.

Using iwspy

You use iwspy to get statistics from specific wireless nodes. With iwspy, you can list the addresses associated with a wireless network interface and get link-quality information for each. The syntax is as follows:

```
iwspy interface [+] DNSNAME | IPADDR | HWADDR [...]
iwspy interface off
```

Let's look at each one of the parameters.

- ✔ **DNSNAME | IPADDR:** Use this parameter to set an IP address or DNS name (using the name resolver).

- ✔ **HWADDR:** Use this parameter to set a hardware (MAC) address.

- ✔ **Plus sign (+):** Use this parameter to add a new set of addresses to the end of the current.

- ✔ **off:** Use this parameter to remove the current list of addresses and to disable the spy functionality.

Using Wavemon

Wavemon is an ncurses-based monitor for wireless devices that polls /proc/net/wireless many times per second. It allows you to watch the signal and noise levels, packet statistics, device configuration, and network parameters of your wireless network hardware. So far, Wavemon has been tested only with the Lucent ORiNOCO series of cards, although it should work (with varying features) with all wireless cards supported by the wireless kernel extensions written by Jean Tourrilhes. You can find Jean's "Wireless Tools for Linux" Web page at www.hpl.hp.com/personal/Jean_Tourrilhes/Linux/Tools.html.

Wavemon continuously updates the statistics. While looking at the statistics, you can press F2 to bring up the Level Histogram. This display gives you a running history of the level of connectivity.

Because Wavemon uses a terminal session, you can simultaneously run more than one instance. You could use each instance to monitor a different link simultaneously.

Wavemon is available from www.janmorgenstern.de/wavemon-current.tar.gz.

Using Wscan

Wscan is a UNIX/X-based link-monitoring application intended for Lucent cards, Linux/x86, Linux/iPaq, or FreeBSD.

The application has two windows. One shows the signal strength. The other window shows details (including ESSID, signal strength, quality, and noise) on a source you select from the signal strength window.

Wscan is available from `www.handhelds.org/download/packages/wscan`.

Using Wmap

Wmap is a tool for creating log files about the reachability of wireless access points with signal strength and GPS coordinates.

Wmap is available from `www.datenspuren.org/wmap`.

Using XNetworkStrength

XNetworkStrength shows signal strength. It's a small application (10.5 KB), is extremely fast, and uses only the X11 API. Oh, and it's free.

XNetworkStrength is available from `http://gabriel.bigdam.net/home/xnetstrength`.

Using Wimon

Wimon is a curses-based wireless connection monitor that shows a real-time graph of a wireless connection's status. It runs on NetBSD, FreeBSD, and OpenBSD. Following is the syntax for Wimon.

```
wimon -i <iface> [-s <scale>] [-d delay in microsec] [-w]
```

Wimon is available from `http://imil.net/wimon`.

Other link monitors

We cover a few tools for monitoring the link quality, but the list of potential tools is long. Following is a list of other link monitors:

- **aphunter** (www.math.ucla.edu/~jimc/mathnet_d/download.html): Link monitor and site survey tool.

- **E-Wireless** (www.bitshift.org/wireless.shtml): Enlightenment link monitor.

- **Gkrellm wireless plug-in** (http://gkrellm.luon.net/gkrellm wireless.phtml): GKrellM monitoring system plug-in.

- **Gnome Wireless Applet** (http://freshmeat.net/projects/gwifi applet): Gnome link monitor.

- **Gtk-Womitor** (www.larsen-b.com/Article/174.html): Applet that shows signal strength.

- **GWireless** (http://gwifiapplet.sourceforge.net): Yet another Gnome link monitor.

- **Kifi** (http://kifi.staticmethod.net): KDE link monitor.

- **KOrinoco** (http://korinoco.sourceforge.net): ORiNOCO-specific link monitor.

- **KWaveControl** (http://kwavecontrol.sourceforge.net): KDE link monitor.

- **KWiFiManager** (http://kwifimanager.sourceforge.net): KDE link monitor and successor to KOrinoco.

- **Mobydik.tk** (www.cavone.com/services/mobydik_tk.aspx): TCL link monitor.

- **NetworkControl** (www.arachnoid.com/NetworkControl/index.html): Monitor interfaces.

- **NetworkManager** (http://people.redhat.com/dcbw/Network Manager): Red Hat/Fedora link monitor.

- **QWireless** (www.uv-ac.de/qwireless): iPaq/Zaurus WLAN analyzer.

- **WaveSelect** (www.kde-apps.org/content/show.php?content=19152): Another KDE link monitor.

- **wmifinfo** (www.zevv.nl/wmifinfo): Applet to display available interface information.

- **WMWave** (www.schuermann.org/~dockapps): Window Maker link monitor.

- **WmWiFi** (http://wmwifi.digitalssg.net/?sec=1): Wireless Monitor for Window Maker.

- **xosview** (http://open-linux.de/index.html.en): Xosview modification to monitor link quality.

Of course, we should mention that the utility that comes with your wireless NIC usually has a link monitor. This is a low-cost, low-fuss solution.

If you have the budget, you might want to consider using a spectrum analyzer like the ones offered by Anritsu (`www.anritsu.co.jp/E/Products/Appli/Wlan`) or Rohde & Schwarz (`www.rohde-schwarz.com`). However, some freeware spectrum analyzers are available — for example, the Waterfall Spectrum Analyzer (`http://freshmeat.net/projects/waterfallspectrum analyzer`). A RF Spectrum Analyzer is a device that receives a chosen range of signals, in our case 2.4 GHz and 5 GHz, and displays the relative signal strength on a logarithmic display, usually a cathode ray oscilloscope.

Network Physical Security Countermeasures

Radio waves travel. This means that crackers don't need to physically attach to your network. Most likely you have locks on your doors. You might even have an alarm system to protect your physical perimeter. Unfortunately, the radio waves don't respect your perimeter security measures. Consequently, you need to walk your perimeter whether you're an individual wanting to protect your access point or a large organization wanting to protect its wired network. While walking the perimeter, monitor the quality of the signal using the tools discussed in this chapter. When you find the signal in places where you don't want it, then turn down the power or move the access point to shape the cell shape.

Other than checking for leakage, you can monitor access points for unauthorized clients.

Checking for unauthorized users

Most access points allow you to view either the DHCP clients or the cache of MAC addresses. This is a good feature for a small network. You can review the cache from time to time to make sure that only your clients are using the access point. If you have only five clients, but you see six MAC addresses, then it just doesn't add up. After you figure out the one that doesn't belong, you can use MAC filtering to block that client.

For a large network, this feature is not very useful. Keeping track of all the MAC addresses in your organization is too difficult. As well, someone running a packet analyzer or sniffer could grab packets and get legitimate MAC addresses. A hacker could then use a MAC address changer like SMAC (`www.klcconsulting.net/smac`), which allows him to set the hardware or MAC address for any interface, say your wireless adapter or Ethernet network interface card (NIC). Figure 6-1 shows the SMAC interface. All you do is put in the hardware address you want and restart the system (or simply disable and re-enable your NIC). Your interface will have the new hardware address.

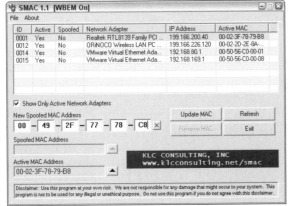

Figure 6-1:
SMAC
interface.

An organization can do any number of things to limit its exposure from the escaping radio waves. The controls are not really technical but rather commonsense. For example, you can change your antenna type.

Antenna type

When placing your access point, you must understand the radiation pattern of the antenna type you choose. The type of antenna you choose directly affects your network's performance, as well as its security.

Before you purchase your antenna, try to obtain a sample radiation pattern. Most antenna vendors supply the specifications for their equipment. You can see a representative radiation pattern and specification for a SuperPass 8 dBi 2.4 GHz antenna at www.superpass.com/SPDG160.html. You can use the specification to determine how far a signal may travel from a particular antenna before becoming unusable; it's just a matter of mathematics.

By understanding the radiation pattern of the antenna you choose, you can do RF signal shaping to "directionalize" the RF signals emitted from your access point. You could switch from an omnidirectional antenna to a semidirectional antenna to control the radiation pattern. Remember, not controlling your signal is equivalent to pulling your UTP cable to the parking lot and letting anyone use it.

Four basic types of antennas are commonly used in 802.11 wireless networks:

 ✔ Parabolic grid

 ✔ Yagi

 ✔ Dipole

 ✔ Omnidirectional

These are discussed in greater detail in the following sections. Figures 6-2 through 6-5 are simplistic depictions of the radiation patterns for the four types of antennae. Each antenna has a unique radiation pattern determined by its construction. We are limited by the print medium, so remember that the radiation pattern is three-dimensional. You may have trouble picturing this; picture a directional antenna as a conical pattern of coverage that radiates in the direction that you point the antenna, while an omnidirectional antenna's pattern of coverage is shaped more like a doughnut around the antenna.

Parabolic grid

Parabolic grid antennae are primarily used for site-to-site applications. A parabolic grid antenna may look like a satellite TV dish or like a wire grid without a solid central core. The parabolic antenna is a unidirectional antenna, meaning that it transmits in one specific direction — the direction that you point the antenna. Figure 6-2 depicts the radiation pattern of a parabolic grid antenna.

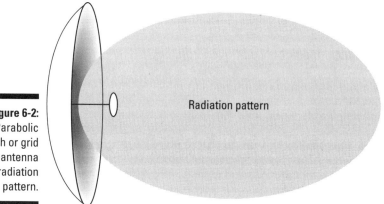

Figure 6-2: Parabolic dish or grid antenna radiation pattern.

Radiation pattern

Yagi

A yagi antenna focuses the beam, but not as much as the parabolic antenna. It's suitable for site-to-site applications in which the distance does not require a parabolic grid. Like the parabolic antenna, a yagi antenna is unidirectional. Figure 6-3 depicts the radiation pattern of a yagi antenna.

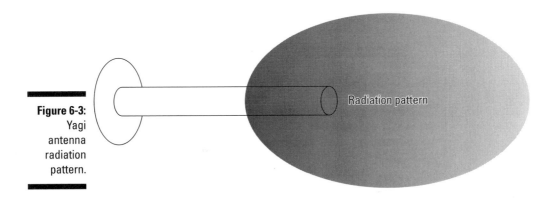

Dipole

A dipole is a bidirectional antenna, hence the use of the suffix *di-*. You generally use a dipole antenna to support client connections rather than site-to-site applications. Figure 6-4 depicts the radiation pattern from the dipole antenna in two directions outward.

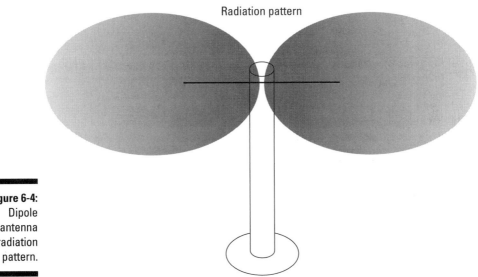

Omnidirectional

An omnidirectional antenna is one that radiates in all directions, losing power as the distance increases. Figure 6-5 depicts the radiation pattern extending in all directions outward. Many wireless base stations come with a small omnidirectional antenna.

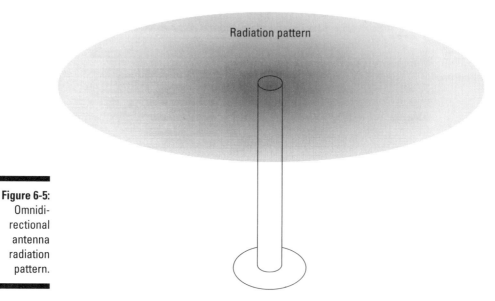

Radiation pattern

Figure 6-5:
Omnidi-
rectional
antenna
radiation
pattern.

Adjusting your signal strength

If you find your signals are bleeding over beyond your perimeter, the first thing you need to do is to reduce the signal strength by adjusting the power settings on your access point. By doing this, you can do some cell sizing and cell shaping. Any access point not meant for the mass home market should allow you to tweak the power. Consider reducing the power of your access point to weaken the signal so that it travels a shorter distance and *doesn't* go where you *don't* want it. If you have a Cisco Aironet 340, for example, you can drop the power output from 30 mW to 5 mW.

If you adjust the power and the signal is still too strong, you need to introduce some loss through the use of an attenuator. You can pick up an attenuator at any good electronics store or find them on the Web. Coaxicom (www.coaxicom.com) is a good place to look for attenuators.

Finally, if changing the antenna type or reducing the power doesn't work, try something simple like moving the access point around your floor. If you have the antenna near an outside wall, the signal will likely seep outside the building. Moving it to an interior location may result in the signal being unusable outside the exterior wall. (You should have found this out when you did your site survey. If your organization did not perform a formal site survey, you might want to get a wireless networking book and read up on site surveys. *Wireless Networks For Dummies* (Wiley) provides everything you need to know to perform your site survey and discusses software to do RF prediction.)

You can also change radiation patterns of your wireless network devices by changing the location of your access points and antennas in relation to large metal objects such as filing cabinets and metal doors. Because radio waves (especially very high frequencies and microwave signals) are easily reflected by metal objects, shadowing, blocking, and reflection of radio signals can be accomplished by the placement of your access points and antennas.

Chapter 7

Hacking Wireless Clients

*T*his book focuses mostly on attacks against wireless *networks* as a whole — that is, 802.11-based attacks against encryption, authentication, and other protocol weaknesses. However, it's important not to forget the reason we have and use networks in the first place — our client systems. When we say *client systems,* we mean workstations, servers, and even APs that are reachable across the wireless network. If wireless networks are accessible to unauthorized people outside your organization, a lot of information can be gleaned from wireless clients. Many hacks don't even require the attacker to be authenticated to the client systems.

When you start poking around on your network, you may be surprised at how many of your wireless clients have security vulnerabilities and just what information they can reveal to attackers. That's why performing security scans on your wireless clients can be so important: It can show you what the bad guys can see if they ever are able to break through your airwaves and gain access to your network hosts.

Think like a hacker — build a mental picture of what's available to be hacked and determine methods to go about exploiting the vulnerabilities.

This chapter shows you how to test for some common wireless-client vulnerabilities. We start with how to scope out wireless hosts on the network and then move on to vulnerabilities that are specific to wireless hosts. We also outline some practical countermeasures, so you can make sure that your systems are secure.

For an in-depth look at detailed vulnerabilities across various wireless client operating systems, e-mail, malware, and more, be sure to check out Kevin's book *Hacking For Dummies* (Wiley).

What Can Happen

If your wireless systems are breached and a hacker is able to obtain access to your internal computers, several bad things can happen. First off, the attacker can gather information about your systems and their configuration, which can lead to further attacks. Such information includes:

- ✔ Open ports and available services
- ✔ Weak passwords
- ✔ WEP keys that are stored locally and not properly secured
- ✔ Acceptable usage policies and banner page information
- ✔ Operating system, application, and firmware versions returned via banners, error messages, or unique system fingerprints
- ✔ Operating system and application configuration information

The exposure of this information can lead to bigger problems such as:

- ✔ Leakage of confidential information, including files being copied and private information such as social security numbers and credit-card numbers being stolen
- ✔ Passwords being cracked and used to carry out other attacks
- ✔ Servers being shut down, rebooted, or taken completely offline
- ✔ Entire databases being copied, corrupted, or deleted

 If you discover a surprising number of vulnerabilities in your wireless APs, workstations, and servers (and you likely will), don't panic. Start by addressing the issues with your most critical systems that will give you the highest payoff once secured.

Although wireless networks are used as a niche solution for many organizations, others are completely dependent on them for all their network connectivity. Either way, wireless networks can serve as an entry point to your workstations, servers, and other wired systems. Therefore, if your wireless client security vulnerabilities aren't addressed and managed properly, they can pose unnecessary risks to the entire network and organization.

Probing for Pleasure

There's a method to the madness of ethical hacking, and testing wireless client security is no different. This involves the ethical hacking steps we discussed in Chapter 3:

- ✔ Gathering public information such as domain names and IP addresses that can serve as a good starting point
- ✔ Mapping your network to get a general idea of the layout
- ✔ Scanning your systems to see which devices are active and communicating
- ✔ Determining what services are running
- ✔ Looking for specific vulnerabilities
- ✔ Penetrating the system to finish things off

These steps are discussed in greater detail in the sections that follow.

Even without poking and prodding your wireless systems further, you may already have vulnerabilities, so don't discount what you've found just because you've gotten this far in the ethical-hacking process. This includes vulnerabilities such as default SSIDs, WEP not being enabled, and critical servers being accessible through the wireless network.

Port scanning

A *port scanner* is a software tool that scans the network to see who is accessing the network and what applications are running. Using a port scanner can help you identify the following:

- ✔ Active hosts on the network
- ✔ IP addresses of the hosts discovered

    ✔ MAC addresses of the hosts found

    ✔ Services or applications that the hosts may be running

    ✔ Unauthorized hosts or applications

The big-picture view from port scanners often uncovers security issues that may otherwise go unnoticed. Port scanners are easy to use and can test systems regardless of what operating systems and applications are running. The tests can be performed very quickly without having to touch individual network hosts, which would be a real pain otherwise.

A good way to get a quick overview of which systems are alive and kicking on the network is to perform a *ping sweep.* A ping sweep is when you send out ping requests (that is, ICMP echo requests) and see if echo replies are received back. Free port scanner programs such as Foundstone's SuperScan (`www.foundstone.com/resources/proddesc/superscan.htm`) and SoftPerfect's Network Scanner (`www.softperfect.com/products/network scanner`), as shown in Figure 7-1, often have ping sweep capabilities built in, and are all you need to get started.

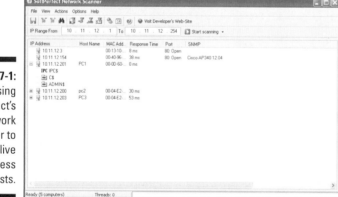

Figure 7-1:
Using
SoftPerfect's
Network
Scanner to
find live
wireless
hosts.

Network Scanner also performs ARP lookups and displays each host's MAC address. This capability is especially handy when testing wireless network security — practically every other tool refers to wireless hosts by their MAC address (or BSSID). The MAC address enables you to easily match up systems you find using NetStumbler, Kismet, or your favorite wireless sniffer with their actual hostnames and IP addresses without having to perform cumbersome reverse-ARP lookups.

Looking for open ports to see what's listening and running on each system is also important. SuperScan is a great tool to use for this because it's easy to use, and it's free! Kevin's partial to SuperScan version 3, as shown in Figure 7-2, because he's been using it for so long, and it simply works.

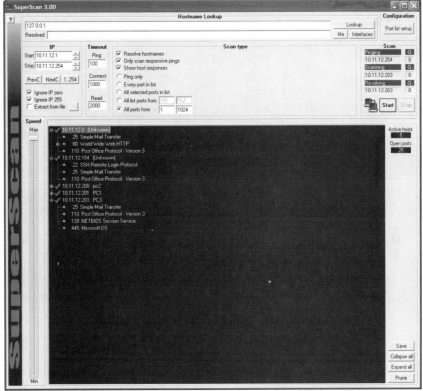

Figure 7-2:
Using
Found-
stone's
SuperScan
to probe
wireless
systems for
open ports.

When performing your network scans, be sure to look for commonly hacked ports, such as those in Table 7-1. Hackers look for these ports, too.

Table 7-1	Commonly Hacked Wireless Network Ports	
Port Numbers	*Service*	*Protocols*
20	FTP data (File Transfer Protocol)	TCP
21	FTP control	TCP
22	SSH	TCP
23	Telnet	TCP
25	SMTP (Simple Mail Transfer Protocol)	TCP
53	DNS (Domain Name System)	UDP
80	HTTP (HyperText Transfer Protocol)	TCP

(continued)

Table 7-1 *(continued)*

Port Numbers	Service	Protocols
110	POP3 (Post Office Protocol version 3)	TCP
135	RPC/DCE end point mapper for Microsoft networks	TCP, UDP
137, 138, 139	NetBIOS over TCP/IP	TCP, UDP
161	SNMP (Simple Network Management Protocol)	TCP, UDP
443	HTTPS (HTTP over SSL)	TCP
512, 513, 514	Berkeley *r* commands (such as rsh, rexec, and rlogin)	TCP
1433	Microsoft SQL Server	TCP, UDP
1434	Microsoft SQL Monitor	TCP, UDP
3389	Windows Terminal Server	TCP

Notice in Figure 7-2 that TCP port 22 (SSH) is open on host 10.11.12.154, which is the access point (AP) on the network. To find out if it's an AP, you can run a NetStumbler, Wellenreiter, or another wireless discovery tool and match the MAC address found there with what Network Scanner finds.

After performing a generic sweep of the network, you can dig deeper into specific hosts you've found. Hmmmm — perhaps a few SSH login attempts on the AP in Figure 7-2 above could get us somewhere?

Using VPNMonitor

A common security measure used to protect wireless data in transit — above and beyond WEP — is to use a Virtual Private Network (VPN). If you installed or manage the VPNs in your organization, you probably know which clients are using them. Then again, if your network is fairly complex, you may not. A free tool you can use to discover whether or not VPNs are being used where they're supposed to be — and thus, whether or not policy is being adhered to — is VPNMonitor (http://sourceforge.net/projects/vpnmonitor).

VPNMonitor sniffs the network and looks for specific signatures belonging to IPsec, PPTP, SSH, and HTTPS traffic. Figure 7-3 shows a basic capture of some VPN traffic, including an SSH connection to the AP at 10.11.12.154, which is denoted by a red line in VPNMonitor.

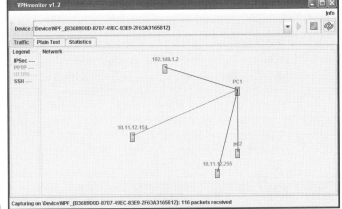

Figure 7-3:
Using
VPNMonitor
to look for
VPN traffic
on the
network.

Wireless networks use a shared communications medium, so it's trivial to capture this type of traffic off the airwaves. However, if you'd like to use VPNMonitor to check for VPN traffic going across your wired network, you can either plug in to a monitor or span port on an Ethernet switch or use a tool such as Ettercap to perform ARP poisoning to make your switch(es) act like a hub. Just be careful because a tool such as Ettercap can take your entire network down if your switch is overly sensitive to ARP poisoning. We cover Ettercap and ARP poisoning in Chapter 12.

Looking for General Client Vulnerabilities

After you find out which wireless systems are alive on your network, you can take your testing a step further and see which vulnerabilities really stand out. There are various freeware, open source, and commercial tools to help you along with your efforts including:

- ✔ **LanSpy (www.lantricks.com):** LanSpy is a Windows-based freeware tool for enumerating Windows systems.

- ✔ **Amap (http://thc.org/thc-amap):** Amap is an open source Linux- and Windows-based application mapping tool.

- ✔ **Nessus (www.nessus.org):** This is an open source network and OS vulnerability-assessment tool that runs on Linux and Windows.

- ✔ **GFI LANguard Network Security Scanner (www.gfi.com/lannetscan):** This is a Windows-based commercial tool for performing network and OS vulnerability assessments.

- ✔ **QualysGuard (www.qualys.com):** QualysGuard is an application service, provider-based commercial tool for performing network and OS vulnerability assessments.

Keep in mind that you'll need more than one security-testing tool. No single tool can do everything.

The presence of these vulnerabilities is why it's so important to run personal firewall and IPS software, such as BlackICE for Windows (`http://blackice.iss.net`) and GNOME-Lokkit (`www.gnome.org`), for Linux systems.

Again, we want to remind you that the tests and vulnerabilities we outline here are just the tip of the iceberg, so check out *Hacking For Dummies* for more details.

Common AP weaknesses

Your wireless APs are wireless clients with operating systems and insecure programs just like any other computer. One of the best ways to check for AP vulnerabilities is to use an all-in-one vulnerability-assessment program, such as Nessus, LANguard Network Security Scanner, or QualysGuard. (QualysGuard is shown in Figure 7-4.)

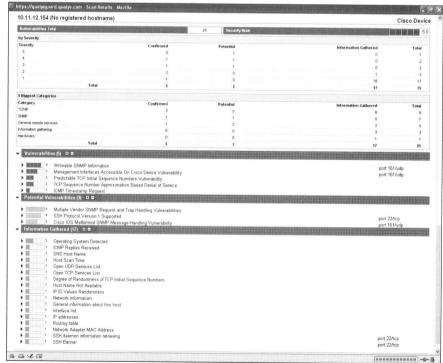

Figure 7-4:
Using Qualys-Guard to dig out vulnera-bilities in a Cisco AP.

Notice in Figure 7-4 how the AP contains common vulnerabilities such as:

- SNMP issues (Vulnerabilities section)
- Weak version of SSH (Potential Vulnerabilities section)
- Open UDP and TCP services (Information Gathered section)
- SSH banner information (Information Gathered section)

Many of these vulnerabilities are not critical, but at least these vulnerabilities need to be addressed because they can likely lead to further AP and network compromise.

Linux application mapping

When it comes to Linux client security, a common attack is against applications with known security vulnerabilities. These applications include FTP, telnet, sendmail, and Apache. Vulnerabilites in these applications can be determined through *application mapping*. A nice — and regularly maintained — tool you can use for application mapping is Amap.

Amap is a very fast application scanner that can grab banners that include version information and even can detect applications that are configured to run on nonstandard ports, such as when Apache is running on port 1711 instead of its default 80. The output of an Amap scan run against a local host is shown in Figure 7-5.

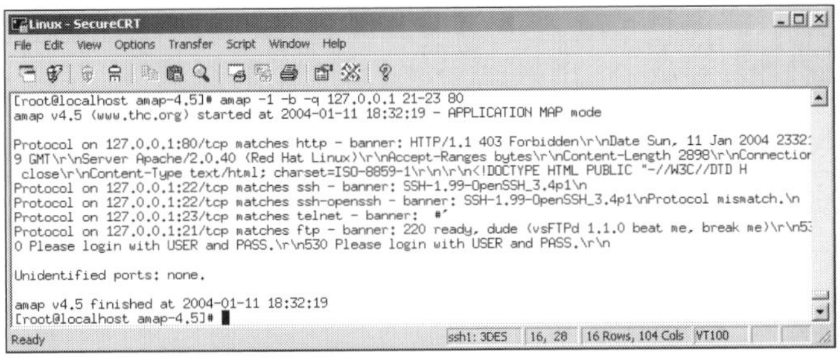

Figure 7-5: Using Amap to check application versions.

Notice that SSH, telnet, and FTP servers were discovered. As is the case here, by perusing the support sites of the applications you discover with Amap, you'll likely find that they've been updated with newer versions to fix various security problems.

Windows null sessions

A well-known vulnerability within Windows can map an anonymous connection *(null session)* to a hidden share called IPC$ (interprocess communication). This attack method can be used to gather Windows information such as user IDs and share names and even allow an attacker to edit parts of the remote computer's registry.

Windows XP and Server 2003 don't allow null session connections by default, but Windows 2000 and NT systems do, so to protect yourself don't forget to test *all* your wireless clients.

Mapping

To map a null session, follow these steps for each Windows computer to which you want to map a null session:

1. **At a command prompt from your test computer, enter the following command. Format the basic net command like this:**

   ```
   net use \\host_name_or_IP_address\ipc$ "" "/user:"
   ```

   The net command to map null sessions requires these parameters:

   - net (the built-in Windows *network* command) followed by the use command
   - IP address of the system to which you want to map a null connection

   - A blank password and username

     The blanks are why it's called a *null* connection.

2. **Press Enter to make the connection.**

   Figure 7-6 shows an example of the complete command when mapping a null session. After you map the null session, you should see the message The command completed successfully.

Figure 7-6:
Mapping
a null
session to a
Windows
2000 server.

To confirm that the sessions are mapped, enter this command at the command prompt:

```
net use
```

As shown in Figure 7-6, you should see the mappings to the IPC$ share on each computer to which you successfully made a null session connection.

Gleaning information

With a null session connection, you can use other utilities to remotely gather critical Windows information. Dozens of tools can gather this type of information. You — like a hacker — can take the output of these enumeration programs and attempt (as an unauthorized user) to try to glean information in the following manners:

- ✔ **Cracking the passwords of the users found.** Be sure to check out *Hacking For Dummies* for a detailed look at password attacks. This chapter can also be downloaded for free at `http://searchsecurity.tech target.com/searchSecurity/downloads/HackingforDummiesCh07.pdf`.

- ✔ **Mapping drives to the network shares to gain access to files, databases, and more.**

You can use Foundstone's SuperScan version 4 to perform automated null session connections and Windows system enumeration as shown in Figure 7-7.

Foundstone's SuperScan version 4 can be found at `www.foundstone.com/resources/proddesc/superscan.htm`.

Keep in mind that Windows XP and Server 2003 are much more secure than their predecessors against such system enumeration vulnerabilities and null session attacks. If such systems are in their default configuration, it should be secure; however, you should still perform these tests against your Windows XP and Server 2003 systems to be sure.

Snooping for Windows shares

Windows *shares* — the available network drives that show up when browsing the network in My Network Places — are often misconfigured, allowing more people to have access to them than necessary. How this works (that is, the default share permission) depends on the Windows system version, as follows:

✔ **Windows NT and 2000:** When creating shares, the group Everyone is given Full Control access in the share by default for all files to browse, read, and write files. Anyone who maps to the IPC$ connection with a null session is automatically made part of the Everyone group! This means that remote hackers can automatically gain browse, read, and write access to a Windows NT or 2000 server if they establish a null session.

✔ **Windows XP and 2003 Server:** The Everyone group is given only Read access to shares. This is definitely an improvement over the defaults in Windows 2000 and NT, but it's not the best setting for the utmost security. You may not even want the Everyone group to have Read access to a share.

Tools such as Legion (`http://packetstormsecurity.nl/groups/rhino9/legionv21.zip`), LanSpy, and LANguard Network Security Scanner can enumerate shares on Windows systems. Imagine the fun a hacker could have with the shares found in the results shown in Figure 7-8!

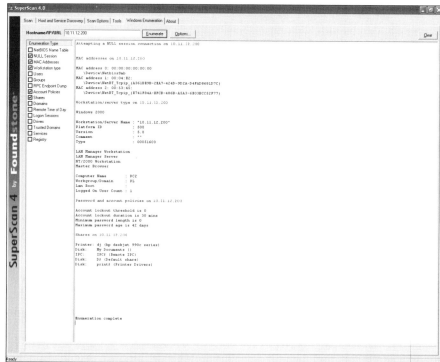

Figure 7-7:
Using Super-Scan to auto-matically create a null session and enumerate a Windows host.

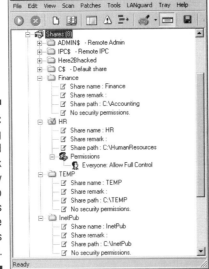

Figure 7-8:
Using
LANguard
Network
Security
Scanner to
find shares
on a remote
Windows
system.

Ferreting Out WEP Keys

Many client vulnerabilities are specific to wireless networks. Standard security tools aren't likely to discover such vulnerabilities. To find these weaknesses, you can use hacking tools that have been created to look for wireless-network vulnerabilities. We discuss such tools below.

Some wireless-specific vulnerabilities require physical access to the computer. It's easy to become complacent and believe that wireless clients are safe because of this physical security requirement, but laptops are lost and stolen quite often, so it's not unreasonable to believe this could occur — especially if users don't report their wireless NICs or laptops stolen. Some vulnerabilities, such as the ORiNOCO WEP key vulnerability, can be exploited by an attacker connecting to the remote computer's registry!

One serious vulnerability affects wireless clients who use the ORiNOCO wireless card. Older versions of the ORiNOCO Client Manager software stores encrypted WEP keys in the Windows registry — even for multiple networks — as shown in Figure 7-9.

You can crack the key by using the Lucent ORiNOCO Registry Encryption/ Decryption program found at `www.cqure.net/tools.jsp?id=3`. Make sure that you use the `-d` command line switch and put quotes around the encrypted key, as shown in Figure 7-10. This program comes in handy if you forget your key, but it can also be used against you.

Figure 7-9:
Encrypted
WEP key
of an
ORiNOCO
wireless
card stored
in the
Windows
Registry.

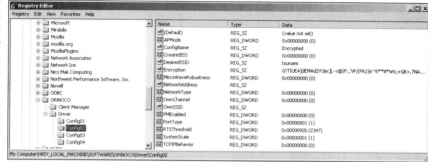

Figure 7-10:
Cracking a
WEP key
stored in the
Windows
registry with
the Lucent
ORiNOCO
Registry
Encryption/
Decryption
program.

If hackers are able to gain remote access to a wireless client through the Connect Network Registry in the Windows Registry editing tool, regedit, they can obtain these keys, crack them, and be on your network in a jiffy.

Wireless NICs from Dell, Intel, and others have all been affected by WEP key storage vulnerabilities — some of which not only store WEP keys in the Windows registry but also store them in plain text!

To find other wireless-specific client vulnerabilities, enter *WEP* into the following vulnerability search engines and compare the results to the wireless hardware and software you may be running.

  ✔ US-CERT Vulnerability Notes Database (www.kb.cert.org/vuls)

  ✔ NIST ICAT Metabase (http://icat.nist.gov/icat.cfm)

  ✔ Common Vulnerabilities and Exposures (http://cve.mitre.org/cve)

Although most of these vulnerabilities are a few years old, you just may find a few weaknesses you weren't expecting.

Wireless Client Countermeasures

Securing all your wireless clients can be quite a task, but you can do some things to keep your systems secure without having to spend a lot of money or effort. At a minimum, ensure the following countermeasures are in place:

✔ Secure your Linux and Windows operating systems.

You can find a ton of great Internet resources for doing this including:

- The Center for Internet Security Benchmark and Scoring Tool for Linux (www.cisecurity.org/bench_linux.html)

- SANS Securing Linux-A Survival Guide for Linux Security (https://store.sans.org/store_item.php?item=83)

- Bastille Linux (www.bastille-linux.org)

- The Center for Internet Security Benchmark and Scoring Tool for Windows 2000, XP, and 2003 (www.cisecurity.org/bench_win 2000.html)

- SANS Securing Windows 2000: Step-by-Step (https://store. sans.org/store_item.php?item=22)

- Microsoft Threats and Countermeasures Guide (www.microsoft. com/technet/Security/topics/hardsys/tcg/tcgch00.mspx)

Also, check out *Hacking For Dummies* and *Network Security For Dummies* for good information on this subject.

✔ Prevent null sessions. You can

- Upgrade your Windows operating systems to XP and Server 2003.

- Block NetBIOS by preventing TCP ports 139 and 445 from passing through your firewall(s).

- Disable File and Print Sharing for Microsoft Networks in the Properties tab of the machine's network connection.

- Create a new DWORD registry key in HKEY_LOCAL_MACHINE\ SYSTEM\CurrentControlSet\Control\LSA called Restrict-Anonymous=1 in the registry for your Windows NT and 2000 systems or setting *Do Not Allow Enumeration of SAM Accounts and Shares* or *No Access without Explicit Anonymous Permissions* in the local security policy or group policy.

The *No Access without Explicit Anonymous Permissions* security setting is not without drawbacks. High security creates problems for domain controller communication and network browsing and the high security setting isn't available in Windows NT.

✔ Install (and require) personal firewall software for every wireless computer.

✔ Disable unnecessary services and protocols on your APs.

✔ Apply the latest firmware patches for your APs and wireless NICs as well as for your client management software.

✔ Regularly perform vulnerability assessments on your wireless workstations as well as your other network hosts.

✔ Apply the latest vendor security patches and enforce strong user passwords.

✔ Use antivirus software *and* antispyware software.

Chapter 8

Discovering Default Settings

· ·

In This Chapter

▶ Collecting information using a sniffer

▶ Grabbing and cracking passwords

▶ Gathering IP addresses

▶ Gathering SSIDs

▶ Protecting yourself

· ·

A first step in testing your wireless network is to glean as much information as you can from "normal" operations. This chapter will introduce you to tools that you can use to look for default settings, sniff traffic, grab passwords, find IP addresses, and discern SSIDs. All information that you can use to further test the security of your wireless network.

Collecting Information

Because your data is traversing the air, anyone with the right tools can sniff the data. In Chapter 2, we introduced you to network or packet analyzers, popularly named sniffers. When it comes to sniffers, you can spend money and buy tools like AiroPeek or CommView for WiFi, or you can save your coin and use some of the excellent free tools exemplified by AirTraf or Ethereal. The next few sections discuss these tools and more.

Are you for Ethereal?

Ethereal, released under the open source license, has many features and compares favorably with commercial products. It works on the UNIX/Linux and Windows platforms, but you must have a pcap library installed. So you may even want to use the tool set in your production environment. Ethereal allows you to capture data from a wired or wireless network. For example, with Ethereal you can read data from IEEE 802.11, Ethernet, Token-Ring, FDDI, and PPP. Not only does it support those media, but it supports 683 protocols. For instance, it can decode 802.11 MGT, 802.11 Radiotap, ARP/RARP, AVS WLAN-CAP, BER, BOOTP/DHCP, CDP, DNS, DOCSIS, EAP, EAPOL, ECHO, Ethernet,

GNUTELLA, GSS-API, HTTP, ICMP, ICQ, IEEE 802.11, IMAP, IP, IRC, ISAKMP, ISDN, KRB5, L2TP, LANMAN, LDAP, LLC,, LSA, LWAPP, LWAPP-CNTL, LWAPP-L3, LWRES, MAPI, NFS, PKCS-1, POP, PPP, PPTP, RPC_NETLOGON, RRAS, RSH, SMB_NETLOGON, SNA, SSH, SSL, Socks, TACACS, TACACS+, TCP, TELNET, TFTP, UDP, VNC, X509AF, X509CE, X509IF, and X509SAT. Fortunately, you can save, print, or filter data.

UNIX/Linux users need the GIMP Toolkit (GTK) for the user interface, whereas the GTK DLLs come bundled with the Windows binary.

You also can use Ethereal as a graphical front-end for packet-capture programs such as Sniffer, `tcpdump`, WinDump, and many other freeware and commercial packet analyzers.

To use Ethereal on a previously created file, you type `tcpdump -w capture.dump` (or WinDump should you wish).

Ethereal is available from `www.ethereal.com`.

This is AirTraf control, you are cleared to sniff

AirTraf was one of the first wireless 802.11b network analyzers. As a wireless sniffer it is a good tool, but does not support wired networks like Ethereal does. It is a passive packet-sniffing tool — it captures and tracks all wireless activity in the coverage area, decodes the frames, and stores the acquired information. AirTraf can record packet count, byte information, related bandwidth, as well as the signal strength of the nodes. You can also run AirTraf in Server Mode, which allows you to have one system that periodically polls other stations to retrieve active wireless data. This is beneficial when you have a large area you want to analyze. You can place AirTraf network analyzers throughout your organization. In this manner, you can consolidate wireless information for your entire organization into a single data store.

AirTraf is Linux open source, and distributed under the GPL license. It is compatible with the 2.4.*x* series of kernels. AirTraf works only with a limited number of wireless adapters. Check the AirTraf Web site to make sure it works with yours.

You can find the freeware AirTraf at `http://airtraf.sourceforge.net/`.

Let me AiroPeek at your data

AiroPeek NX is a Windows-based wireless sniffer that offers some enhanced capabilities, including the ability to detect rogue access, risky device configurations, Denial-of-Service attacks, Man-in-the-Middle attacks, and intrusions. We have used AiroPeek and highly recommend it. There is one drawback to AiroPeek: It is a commercial product. However, after you use it we think you'll agree that it is money well spent. This is one tool we would recommend that you spend your hard-earned money on if you're going to do more than one ethical hack.

AiroPeek NX comes with a Security Audit Template that creates a capture window and then triggers a notification when any packet matches a specifically designed security filter. This allows the administrator to search for applications like Telnet and access points that use default — therefore *not secure* — configurations.

If you are using Network Authentication with protocols such as Telnet and FTP, you can use AiroPeek to look for failed authentications. These failures might represent an attempted access by an unauthorized person. Once you start to look at the data you are collecting, you can dream up all sorts of similar tests using a sniffer or packet analyzer.

You can find AiroPeek NX at www.wildpackets.com.

Another CommView of your data

Another wireless sniffer is CommView for WiFi, which is specific to wireless networks and offers many capabilities besides packet sniffing, such as statistical analysis. By doing statistical analysis, you might find a pattern of unauthorized usage. CommView allows you to grab frames, store the information, and analyze it. CommView for WiFi is a commercial product. You'll find it's not as expensive as AiroPeek but (obviously) more costly than the free Gulpit and Ethereal programs.

When CommView for WiFi is running on your machine, it places your wireless adaptor in passive mode. Your wireless interface can only capture all the packets when it is in passive mode. You will find the installation fairly straightforward since it uses the Windows installer process. Once you install it, you will find many options as shown in Figure 8-1.

You can find Tamosoft CommView for WiFi at www.tamos.com/products/commview/.

You cannot obtain data from an access point using WEP or WPA unless you have the appropriate key. You can add key information to CommView for WiFi by selecting Settings⇨WEP/WPA Keys and then entering the keys in the areas provided.

Start icon

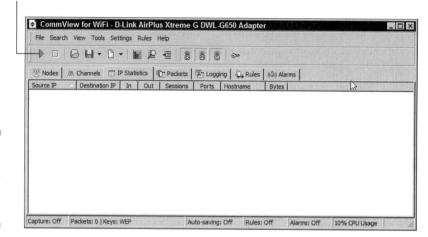

CommView for WiFi - D-Link AirPlus Xtreme G DWL-G650 Adapter

File Search View Tools Settings Rules Help

Nodes | Channels | IP Statistics | Packets | Logging | Rules | Alarms

Source IP | Destination IP | In | Out | Sessions | Ports | Hostname | Bytes

Capture: Off | Packets: 0 | Keys: WEP | Auto-saving: Off | Rules: Off | Alarms: Off | 10% CPU Usage

Figure 8-1:
Viewing the
CommView
for WiFi
main menu.

To start using all the tabs shown in Figure 8-1, you need to begin capturing packets so you can obtain some actual data. After you identify and input the proper keys, you need to start the capture process. Simply follow these steps:

1. **Start the CommView program.**

2. **Click the Start icon.**

   Alternatively, select File⇨Start.

3. **From the Scanner section that appears in the new window, click on Start Scanning.**

   The program will start scanning all channels for wireless signals and display them under the Access Points and Hosts section.

4. **Select one of the networks displayed to produce details about that network under the term Details.**

   The Details are shown in Figure 8-2.

5. **Choose one of the networks and click the Capture button.**

   CommView begins to capture packets.

6. **To view the current bandwidth load for a network, Select View⇨Statistics.**

7. **To run a report, use the Report tab and select either HTML format or comma-delimited format.**

   This report provides a report on overall performance of your network.

8. **Select File⇨Stop Capture to shut down CommView.**

Look through the frames you gather for potentially useful information such as login frames.

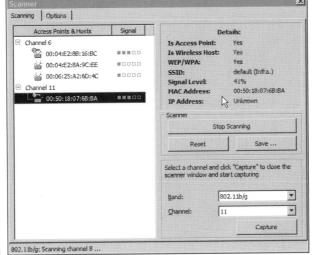

Figure 8-2:
Viewing the
CommView
for WiFi
Scanner
page.

Gulpit

Gulpit is based on Trinux and does not require an operating system. You don't even really need a hard drive to use it. Gulpit boots from a CD-ROM (of course, you must set up your BIOS to boot from your CD-ROM first). Gulpit is released as open source.

Gulpit will turn your laptop with an ORiNOCO 802.11b wireless card (or any OEM clones such as Agere and Proxim) into a packet sniffer for your wired and wireless networks. Gulpit is a *packet gulper.* A packet gulper is nothing more than a really good packet sniffer. Packet sniffers read essentially *all* the information and control structures on a wired or wireless network.

You will find that only certain cards support radio-monitor mode. ORiNOCO cards obviously work. You'll find that Prism II cards generally work. Cards that do not support radio monitor mode will work with `tcpdump` and `tethe real` but not Kismet. You can read the Gulpit documentation to find out what wireless cards it does support.

It sniffs Ethernet frames as well as wireless 802.11b frames. The wireless frames are sniffed in the radio monitor mode so you see just about all the traffic on the air regardless of its protocol.

Starting Gulpit is as easy as the following steps.

1. **Open your laptop CD-ROM bay.**

2. **Power down Windows or Linux or whatever operating system you are using.**

3. **Insert the Gulpit CD-ROM into the drive and close it.**

4. **Power up the computer and watch Gulpit boot.**

   If this is the first time you have used Gulpit, interrupt the boot process and enter the BIOS set-up program. Make sure that your system will boot from the CD-ROM before the hard drive and disable your floppy controller when it is enabled. Obviously the exact method for doing this depends on your hardware manufacturer and the BIOS you are using, but look at the screen and follow the instructions to enter the set-up program. Once set, you won't need to do this again.

   You must disable your floppy in your BIOS settings to use Gulpit.

5. **You will see a Gulpit splash screen with license and credit information.**

   Gulpit will then pause temporarily at a `boot:` prompt. Hit Return (or Enter) at this time — or just wait a few seconds and Gulpit will continue on its own.

Whenever you want to use Gulpit, just put the disc in the drive and turn on the power. When you're finished with Gulpit, remove the Gulpit CD-ROM and reboot. Your system will boot whatever operating system from your hard drive (assuming you have set it up that way).

Gulpit installs itself and a complete Linux 2.4.5 kernel on a ramdisk and executes in RAM. Gulpit has complete PCMCIA support and is ideally run on a laptop computer. Gulpit will not make a mark on your hard drive unless you want to store data there. In that case, Gulpit has support for fat, ntfs (read-only) and vfat as well as minix and ext2 file systems.

Gulpit has three tools for packet sniffing. Each one has its own capabilities and limitations:

- ✔ **Kismet:** Gulpit is set up to start Kismet automatically in "radio monitor" and frequency hopping mode. This will log all the traffic from nearby transmitters. If you don't want to start Kismet, or you wish to operate Kismet in the single channel mode, just hit ctrl-C as the boot process completes and as Kismet starts. This will cleanly shut down Kismet. You can learn more about Kismet in Chapter 10.

- ✔ **Tethereal:** Tethereal collects and decodes a multitude of protocols. Tethereal is the curses (text based) version of Ethereal. Tethereal supports packet capture in the radio monitor mode. Tethereal will sniff wireless as well as wired packets. Ethereal has a nice graphical display for tethereal and Ethereal collected packets as well as those collected by tcpdump and Kismet.

✔ **Tcpdump:** Tcpdump also collects and decodes a multitude of protocols. It is basically like Tethereal but does not work with wireless networks because it does not work in radio monitor mode.

You can find Gulpit at `www.crak.com/gulpit.htm`.

The developer of Gulpit recommends you check out Auditor Linux at `http://new.remote-exploit.org/index.php/Auditor_main` or download it from `http://mirror.switch.ch/ftp/mirror/auditor/auditor-120305-01.iso.zip`. Auditor is also a bootable version of Linux with many of the wireless tools built in.

That's Mognet not magnet

Mognet is a simple, lightweight 802.11b sniffer written in Java and available under the GNU Public License (GPL). It was designed for handheld devices like the iPAQ, but will run on a desktop or laptop. Mognet features real-time capture output, support for all 802.11b generic and frame-specific headers, easy display of frame contents in hexadecimal or ASCII, text mode capture for GUI-less devices, and loading/saving capture sessions in libpcap format.

You can find Mognet at `www.10t3k.net/tools/Wireless/Mognet-1.16.tar.gz`.

Other analyzers

Not fond of any of the programs discussed so far, well don't despair. There are plenty of alternatives. Following is a list of wireless packet analyzers:

✔ **AirMagnet (`www.airmagnet.com/`):** Commercial product

✔ **AirScanner Mobile Sniffer (`http://airscanner.com/downloads/sniffer/sniffer.html`):** Freeware product

✔ **Capsa (`www.colasoft.com/products/capsa/index.php?id=75430g`):** Commercial product

✔ **CENiffer (`www.epiphan.com/products_ceniffer.html`):** Commercial product

✔ **KisMAC (`www.binaervarianz.de/projekte/programmieren/kismac/`):** Freeware product

✔ **Kismet (`www.kismetwireless.net/`):** Freeware product

✔ **LANfielder** (`www.wirelessvalley.com/`): Commercial product

✔ **LinkFerret** (`www.baseband.com/`): Commercial product

✔ **ngrep** (`www.remoteassessment.com/?op=pub_archive_search& query=wireless`): Freeware product

✔ **Observer** (`www.networkinstruments.com/`): Commercial product

✔ **Packetyzer** (`www.networkchemistry.com/`): Commercial product

✔ **Sniffer Netasyst** (`www.sniffer-netasyst.com/`): Commercial product

✔ **Sniffer Wireless** (`www.networkgeneral.com/Products_details. aspx?PrdId=20046178370181`): Commercial product

✔ **SoftPerfect Network Protocol Analyzer** (`www.softperfect.com/ products/networksniffer/`): Commercial product

Should you not find anything above, again don't despair because you can find information about wireless sniffers at either `www.personaltelco.net/ index.cgi/WirelessSniffers`, `www.winnetmag.com/Files/25953/ 25953.pdf`, or `www.blacksheepnetworks.com/security/resources/ wireless-sniffers.html`.

Cracking Passwords

After you are connected at Layer 2, you'll want to sniff some passwords and crack them. There are lots of wonderful tools to do this, and we have selected two of the better ones for you: Cain & Abel, a.k.a. Cain and `dsniff`.

Using Cain & Abel

Cain & Abel is a freeware password recovery tool that runs on a Microsoft platform. It allows easy recovery of various kinds of passwords by sniffing the network, cracking encrypted passwords using Dictionary, Brute-Force and Cryptanalysis attacks, recording VoIP conversations, decoding scrambled passwords, revealing password boxes, uncovering cached passwords and analyzing routing protocols. This tool covers some security weaknesses present in the protocols, authentication methods and caching mechanisms.

Cain & Abel was developed for network administrators, security consultants or professionals, forensic staff, security-software vendors, and professional penetration testers.

Should you use Cain & Abel, be *very* careful. First, you should understand that when using a password cracker, you may violate any number of wiretapping laws or put your organization in a precarious position. If you know passwords are weak and you don't immediately change them, you might have difficulty proving due diligence in a court of law. So ensure that you are on the right side of the law before you touch a key. Second, there is the remote possibility that you could cause damage or the loss of data when using this software or similar tools. These tools intercept packets and may damage these packets. Ensure that you know how the tool works and what it could do — and that good recent backups of system data exist.

The latest version is faster and contains a lot of new features like APR (ARP Poison Routing) that facilitates the sniffing of switched LANs and Man-in-the-Middle attacks. You can use Cain to analyze encrypted protocols such as SSH-1 and HTTPS and to capture credentials from a wide range of authentication mechanisms. It also provides routing protocols authentication monitors and route extractors, dictionary and brute-force crackers for all common hashing algorithms and for several specific authentications, password/hash calculators, cryptanalysis attacks, password decoders, and some not so common utilities related to network and system security. This is indeed the Swiss Army knife for password crackers. Figure 8-3 shows the main window of Cain.

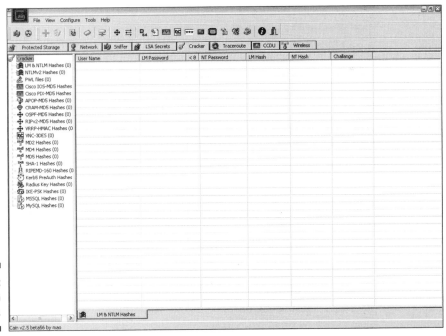

Figure 8-3: Cain main window.

Cain & Abel is actually two different programs. Cain has the following features:

- ✔ **Protected Storage Password Manager:** Reveals locally stored passwords of Outlook, Outlook Express, Outlook Express Identities, Outlook 2002, Internet Explorer, and MSN Explorer.

- ✔ **Credential Manager Password Decoder:** Reveals passwords stored in Enterprise and Local Credential Sets on Windows XP/2003.

- ✔ **LSA Secrets Dumper:** Dumps the contents of the Local Security Authority Secrets.

- ✔ **Dialup Password Decoder:** Reveals passwords stored by Windows "Dial-Up Networking" component.

- ✔ **APR (ARP Poison Routing):** Enables sniffing on switched networks and Man-in-the-Middle attacks.

- ✔ **Route Table Manager:** Provides the same functionality of the Windows tool route.exe with a GUI front-end.

- ✔ **SID Scanner:** Extracts usernames associated with Security Identifiers (SIDs) on a remote system.

- ✔ **Network Enumerator:** Retrieves, where possible, the user names, groups, shares, and services running on a machine.

- ✔ **Service Manager:** Allows you to stop, start, pause, continue, or remove a service.

- ✔ **Sniffer:** Captures passwords, hashes, and authentication information during transmission on the network. Includes several filters for application specific authentications and routing protocols. The VoIP filter enables the capture of voice conversations transmitted with the SIP/RTP protocol saved later as WAV files.

- ✔ **Routing Protocol Monitors:** Monitors messages from various routing protocols (HSRP, VRRP, RIPv1, RIPv2, EIGRP, OSPF) to capture authentications and shared route tables.

- ✔ **Full SSH-1 sessions sniffer for APR (APR-SSH-1):** Allows you to capture all data sent in a HTTPS session on the network.

- ✔ **Full HTTPS sessions sniffer for APR (APR-HTTPS):** Allows you to capture all data sent in a HTTPS session on the network.

- ✔ **Certificates Collector:** Grabs certificates from HTTPS Web sites and prepares them for use by APR-HTTPS.

- ✔ **MAC Address Scanner with OUI fingerprint:** Using OUI fingerprint, makes an informed guess about the device based on the MAC address.

- **Promiscuous-mode Scanner based on ARP packets:** Identifies sniffers and network intrusion detection systems present on the LAN.

- **Wireless Scanner:** Scans for wireless networks signal within range. This feature is based on NetStumbler that we discuss in Chapter 9.

- **Access (9x/2000/XP) Database Passwords Decoder:** Decodes the stored encrypted passwords for Microsoft Access Database files.

- **Base64 Password Decoder:** Decodes Base64 encoded strings.

- **Cisco Type-7 Password Decoder:** Decodes Cisco Type-7 passwords used in router and switches configuration files.

- **VNC Password Decoder:** Decodes encrypted VNC passwords from the registry.

- **Enterprise Manager Password Decoder:** Decodes passwords used by Microsoft SQL Server Enterprise Manager (SQL 7.0 and 2000 supported).

- **Remote Desktop Password Decoder:** Decodes passwords in Remote Desktop Profiles (.RPD files).

- **PWL Cached Password Decoder:** Allows you to view all cached resources and relative passwords in clear text either from locked or unlocked password list files.

- **Password Crackers:** Enables the recovery of clear text passwords scrambled using several hashing or encryption algorithms. All crackers support Dictionary and Brute-Force attacks.

- **Cryptanalysis attacks:** Enables password cracking using the "Faster Cryptanalytic time – memory trade off" method introduced by Philippe Oechslin. This cracking technique uses a set of large tables of pre-calculated encrypted passwords, called Rainbow Tables, to improve the trade-off methods known today and to speed up the recovery of cleartext passwords.

- **NT Hash Dumper + Password History Hashes (works with Syskey enabled):** Retrieves the NT password hash from the SAM file regardless of whether Syskey is enabled or not.

- **Microsoft SQL Server 2000 Password Extractor via ODBC:** Connects to an SQL server via ODBC and extracts all users and passwords from the master database.

- **Box Revealer:** Shows passwords hidden behind asterisks in password dialog boxes.

- **RSA SecurID Token Calculator:** Calculates the RSA key given the tokens .ASC file.

- **Hash Calculator:** Produces the hash values of a given text.

- **TCP/UDP Table Viewer:** Shows the state of local ports (like netstat).

✔ **TCP/UDP/ICMP Traceroute with DNS resolver and WHOIS client:** An improved `traceroute` that can use TCP, UDP and ICMP protocols and provides `whois` client capabilities.

✔ **Cisco Config Downloader/Uploader (SNMP/TFTP):** Downloads or uploads the configuration file from/to a specified Cisco device (IP or hostname) given the SNMP read/write community string.

Abel provides the following features:

✔ Remote Console: Provides a remote system shell on the remote machine.

✔ Remote Route Table Manager: Manages the route table of the remote system.

✔ Remote TCP/UDP Table Viewer: Shows the state of local ports (like net-stat) on the remote system.

✔ Remote NT Hash Dumper + Password History Hashes (works with Syskey enabled): Retrieves the NT password hash from the SAM file regardless of whether Syskey is enabled or not; works on the Abel-side.

✔ Remote LSA Secrets Dumper: Dumps the contents of the Local Security Authority Secrets present on the remote system.

Cain & Abel is a must-have for your ethical-hacking toolkit. You can find Cain & Abel at www.oxid.it/cain.html.

Using dsniff

`dsniff` is a collection of freeware tools for network auditing and penetration testing. `dsniff`, `filesnarf`, `mailsnarf`, `msgsnarf`, `urlsnarf`, and `webspy` passively monitor a network for interesting data (for example, passwords, e-mail, and files). `arpspoof`, `dnsspoof`, and `macof` facilitate the interception of network traffic normally unavailable to an attacker (due to Layer-2 switching). `sshmitm` and `webmitm` implement active monkey-in-the-middle attacks against redirected ssh and https sessions by exploiting weak bindings in ad hoc PKI. The author of `dsniff` tested it himself on OpenBSD, Red Hat Linux, and Solaris, while other individuals have run `dsniff` on FreeBSD, Debian Linux, Slackware Linux, AIX, and HP-UX.

To use `dsniff`, you also will need Berkeley DB, OpenSSL, `libpcap`, `libnet`, and `libnids`. OpenBSD already incorporates the first three packages into the base system, leaving only libnet and libnids as additional dependencies. You can download the latter two from the OpenBSD FTP site. You will find the other OS will require a little more work. `dsniff` is a simple password sniffer that handles authentication information from the following sources:

FTP	IRC
Telnet	AIM
HTTP	CVS
POP	ICQ
NNTP	Napster
IMAP	Citrix ICA
SNMP	Symantec pcAnywhere
LDAP	
Rlogin	NAI Sniffer
NFS	Microsoft SMB
SOCKS	Oracle SQL*Net
X11	

`dsniff` benefits the user because it minimally parses each application protocol, saving only the "interesting" data. This speeds up processing.

`dsniff` is really easy to use. Just start it, and it starts listening on the interface you select for passwords.

Mailsnarf outputs all messages sniffed from SMTP traffic in Berkeley mbox format, suitable for offline browsing with a mail reader, such as `pine`. Urlsnarf outputs all requested URLs sniffed from HTTP traffic in Common Log Format, used by almost all Web servers, suitable for offline post-processing with a Web log-analysis tool, such as analog or `wwwstat`. Webspy sends URLs sniffed from a client to a Netscape browser. Filesnarf outputs NFS, SMB, and AFS. Msgsnarf outputs ICQ, AIM, and IRC.

As well, you can use `dsniff` to perform a monkey-in-the-middle attack using `sshmitm` and `webmitm` to sniff HTTPS and SSH traffic and to capture login information.

You can find `dsniff` at `www.monkey.org/~dugsong/dsniff/`. A Windows port is available from `www.datanerds.net/~mike/dsniff.html`, and a MacOS X port is available at `http://blafasel.org/~floh/ports/dsniff-2.3.osx.tgz`.

Gathering IP Addresses

Crackers want targets, and IP addresses are targets. Also, if the wireless administrator is using MAC filtering, then you'll need to gather some IP addresses. You can `ping` every host on a subnet to get a list of MAC to IP addresses. But this is a tedious task at best. Instead, you can `ping` the broadcast of the subnet, which in turn will `ping` every host on the local subnet. This is what the `arping` tool does for you.

On Windows and some other operating systems, the `arp` command provides access to the local ARP cache. In Windows, for example, typing `arp -a` at the command prompt will display all of the entries in that computer's ARP cache. The ARP cache stores previously resolved hardware or MAC addresses for requested software or IP addresses.

An almost unknown command, `arping` is similar to `ping`, but different in that it works at the Ethernet layer. While `ping` tests the reachability of an IP address, `arping` reports the reachability and round-trip time of an IP address hosted on the local network.

`Arping` works on Linux, FreeBSD, NetBSD, OpenBSD, MacOS X, Solaris, and Windows. Below is the help information for `arping`.

```
Usage: arping [-fqbDUAV] [-c count] [-w timeout] [-I device]
              [-s source]
              destination
          -f : quit on first reply
          -q : be quiet
          -b : keep broadcasting, don't go unicast
          -D : duplicate address detection mode
          -U : Unsolicited ARP mode, update your neighbours
          -A : ARP answer mode, update your neighbours
          -V : print version and exit
          -c count : how many packets to send
          -w timeout : how long to wait for a reply
          -I device : which ethernet device to use (eth0)
          -s source : source ip address
          destination : ask for what ip address
```

There are several ways you can use `arping`. Under normal operation, `arping` displays the Ethernet and IP address of the target as well as the time elapsed between the arp request and the arp reply. Or, you can use the `-U` option to send a broadcast arp and gather IP addresses.

You can find `arping` at `www.habets.pp.se/synscan/programs.php?prog=arping`.

Gathering SSIDs

In the next two chapters, we will show you tools that will assist you in gathering SSIDs. To connect to an access point, you need to know the SSID. Contrary to what some people think, a SSID is not a password, and you should not use it as such.

Using essid_jack

In Chapter 10, we talk about passive and active network discovery. At this point, you just need to know that NetStumbler is an active scanner. Many people suggest that you can defeat those nosy people running NetStumbler out there by disabling SSID broadcast. This indeed does make NetStumbler ineffective; however, you have other options such as Kismet and essid_ jack. You will learn more in Chapter 10 about Kismet, so let's look at essid_ jack now. You can use essid_jack to report the SSID of an access point to you. essid_jack is part of a open source suite of tools labeled air-jack (http://sourceforge.net/projects/airjack/).

The reason essid_jack works even when you disable the SSID is simple: The access point will eventually send the SSID in cleartext when a legitimate client attempts to connect to the access point. Most crackers are impatient, though, and don't want to wait until someone attempts to connect. In effect, essid_ jack impersonates an access point by spoofing its MAC address. It then sends a disassociate frame to the clients causing them to disassociate from the access point. The clients then attempt to reassociate with the access point, and in so doing they transmit an association request with the access point's SSID in cleartext. *Presto!* — essid_jack captures the SSID.

```
# ./essid_jack -h
Essid Jack: Proof of concept so people will stop calling an
    ssid a password.
Usage: ./essid_jack -b <bssid> | [ -d <destination mac> ]
[ -c  <channel number> ] [ -i <interface name> ]

        -b: bssid, the mac address of the access point (e.g.
           00:de:ad:be:ef:00)
        -d: destination mac address, defaults to broadcast
           address.
        -c: channel number (1-14) that the access point is
           on, defaults to current.
        -i: the name of the AirJack interface to use
           (defaults to aj0).
```

Now you know how to use it. So let's try it on a MAC address on channel 6:

```
#./essid_jack -b 00:0c:6e:9f:3f:a6 - c 6
Got it, the essid is (escape characters are c style):
"pdaconsulting"
```

You can find essid_jack by downloading air_jack from http://sourceforge.net/projects/airjack/.

Using SSIDsniff

SSIDsniff is a curses-based tool that allows an intruder to identify, classify, and data-capture wireless networks. The SSIDsniff interface will look familiar if you've ever used the UNIX top utility.

Currently it works under Linux and is distributed under the GPL license. You will need `libpcap` and `curses` or `ncurses` as well. SSIDsniff supports Cisco Aironet and some Prism2 cards.

You can find SSIDsniff at `www.bastard.net/~kos/wifi/ssidsniff-0.40.tar.gz`.

Default-Setting Countermeasures

Okay, even though this chapter introduces you to some very powerful tools, you must not put your head in the sand; just knowing about these tools (and what hackers can do with them) won't make them go away. They are here to stay — and their friends are moving in. Two things we know for sure from the short history of the Internet: These (and other, more insidious tools) proliferate, and they come at you at an ever-increasing pace. Your plan of defense must include ferreting out and trying these tools — as well as their next-generation kid brothers — from here on in. It's an arms race — you must know what the enemy is using, and be prepared to escalate.

The good news is: Some of the countermeasures are decidedly low-tech. There's really no excuse for not implementing them.

Change SSIDs

When you get a new system, you must ensure that you change the default SSID. We know Linksys uses `Linksys` as a default SSID (obvious, much?), and we know others as well. When picking a new SSID — as long as we're talking obvious (but vital) here — don't select one that's easy to guess. Even though the SSID is most emphatically not a password, there is no reason to select an easy-to-guess one.

If you don't know what the default SSID is for a particular access point, you can find it out at one of the following Web sites:

- `www.cirt.net/cgi-bin/passwd.pl`
- `www.phenoelit.de/dpl/dpl.html`
- `http://new.remote-exploit.org/index.php/Wlan_defaults`
- `www.thetechfirm.com/wireless/ssids.htm`

Don't broadcast SSIDs

In this chapter, we showed you that even when you don't broadcast your SSID, others can derive it. But that doesn't mean you shouldn't disable it. When someone roams your neighborhood running NetStumbler, make it more difficult for them. Disable your SSID broadcasting and make them come back running Kismet. You may not have defeated them (yet), but you've at least made things more difficult for them.

Using pong

Older readers probably think pong is a video game. If you are a computer virus researcher or fighter, then you probably think pong is a nasty Trojan. Well, this pong is neither, but rather a tool to check the vulnerability of your wireless access point. If your access point is running vulnerable firmware, pong will give you access to all relevant details such as the admin password, WEP keys, allowed MAC addresses, and more. Should pong work successfully against your network, then you'll need to upgrade your firmware to protect yourself.

Pong is a DOS program and is easy to use, just type `c:\> pong [-r]` in a command shell. The `-r` option provides additional raw output of all received data. When pong finds an access point from the following list, you will get a list of all relevant parameters:

- 4MBO
- Airstation
- D-Link DWL-900AP+
- Linksys
- Melco
- US Robotics
- Wisecom

You can find pong at `http://mobileaccess.de/wlan/index.html?go=technik&sid=`. Praemonitus, praemunitus. (Or for those of you who don't still speak Latin, that's *forewarned, forearmed.*)

Detecting sniffers

At Layer 2, you can run LBL's arpwatch (`www.securityfocus.com/tools/142`) to detect changes in ARP mappings on the local network, such as those caused by `arpspoof` or `macof`.

At Layer 3, you can use a tool such as AiroPeek, CommView for WiFi, or any other programmable sniffer (say, NFR) to look for either the obvious network anomalies or the second-order effects of some of `dsniff`'s active attacks. If you want to learn how to use a packet analyzer for security, try one of Laura Chappell's network analysis or troubleshooting books that you can download for a fee from `www.packet-level.com/books.htm`.

Also, anti-sniffing programs such as l0pht's AntiSniff (`http://packet stormsecurity.nl/sniffers/antisniff/`) can uncover `dsniff`'s passive monitoring tools.

Chapter 9

Wardriving

· ·

In This Chapter

▶ Installing and configuring Network Stumbler

▶ Running NetStumbler

▶ Interpreting the results

▶ Mapping and viewing the results

· ·

*W*hen most people think of wireless security (or the lack of it), they think of someone driving around their neighborhood discovering their access point and trying to connect. This is a striking image: A nerd in a car by himself with his beloved laptop and some arcane software. It's an activity called *wardriving,* and though it seems hostile at first blush, the reality is actually a lot more diverse. In effect, wardriving is an educational opportunity for everyone — especially for ethical hackers. Peter, for example, actually goes wardriving with his teenage daughter. After all, the family that drives together, strives together.

In this chapter, we take our first look at wardriving. To understand this genre of software, we will look at Network Stumbler (a.k.a. NetStumbler). We'll also see how to map the results of your work. In Chapter 10, we discuss other examples of wardriving software, such as Kismet and Wellenreiter.

Introducing Wardriving

The term *wardriving* is derived from the phrase *war dialing.* But it really doesn't involve guns or offensive weapons of any kind. Wardriving is just the term coined for wireless network discovery. Nothing more or less. In Chapter 4, we outlined the tools you need for your wardrive, but all you need to wardrive is some software and a wireless network interface card or adapter. If you really want to get into it, you can add an external antenna to enhance the signal strength of any access points that you find. This enables you to detect these access points at a greater distance than when you were only using the built-in antenna of your wireless NIC alone. You could also add a global positioning system (GPS) to map the latitude and longitude of the networks you find.

Driving a car and watching your computer is a dangerous activity. It may even be illegal. So when you go wardriving, please take someone with you so you can concentrate on the road. We don't want you ending up as "warkill."

Network Stumbler is the application for wardrivers who favor the Windows platform. It runs on Windows 3.9x, Me, 2000, and XP. NetStumbler uses the active scanning method to discover access points; and when it's equipped with a GPS unit, it records the latitude and longitude of any discovered access points. You can later graph the recordings with mapping software.

NetStumbler uses the active scanning method described by the IEEE 802.11 specification to discover wireless networks. It sends multiple probe requests, and records probe responses. You may wonder why this would work, but when you think about it, it makes perfect sense. The developers of this standard made the active scanning option available so clients with multiple unique networks could find all of their available networks.

Once an access point receives a probe request, it typically responds with a *probe-response management frame* containing the network BSSID and the WLAN SSID. Some access points can "cloak" their SSID by responding to probe requests with only a single space for the SSID, forcing users to have prior knowledge of the network SSID before joining the networks. NetStumbler cannot report access points that cloak their SSID. You'll need to skip to Chapter 10 and read about Kismet in that case.

When NetStumbler locates a network, it records the following information:

- **The signal, noise, and signal-to-noise ratio (SNR) of the discovery:** This simplistically can indicate how close you are to the device.

- **The operating channel:** In North America, this is a number between 1 and 11.

- **Basic SSID (BSSID):** This is actually the MAC address of the access point.

- **Service Set Identifier (SSID):** The SSID is a 32-character unique identifier for the network embedded to the header of frames sent over a WLAN.

- **The access point's "nickname":** This is the access point's name.

NetStumbler also has a very useful way of graphing the signal strength of the received APs and other Wi-Fi clients in your surrounding area. This signal strength meter may be used with a directional antenna (such as a cantenna) to help figure out the location of the signal.

We know you want to get started. So let's go.

Installing and Running NetStumbler

Installing NetStumbler is easy. Go to www.netstumbler.com/downloads/ and download it. Run the self-installer and you will have the usual Windows installation-wizard experience. When prompted, select to install all options. You can always delete the shortcut or move things later. If you want to view the files you've installed and their locations, you can do so by clicking the Show details button. You may want to read the README file before you start using NetStumbler.

Running NetStumbler is as easy. Either double-click the Network Stumbler icon on the desktop or choose Network Stumbler under All Programs from the Start menu. Then you see the NetStumbler 0.4.0 splash screen, which shows the adapter, driver information, and MAC address. When NetStumbler starts, it needs no prompting: It immediately attempts to open a new document, locate a wireless adapter and a GPS, and start scanning. NetStumbler starts to capture data in a file labeled YYYYMMDDHHMMSS.ns1, which is based on the date and time of the capture.

Figure 9-1 gives you an example of what you see when you start NetStumbler.

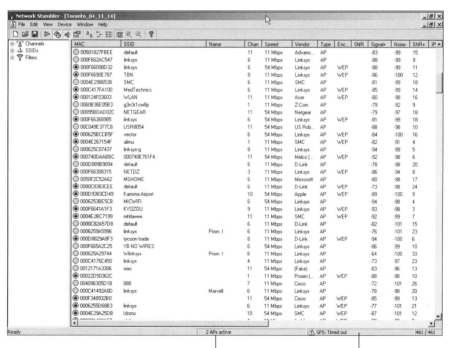

Figure 9-1:
Network
Stumbler
window.

Active access points   Status information

Setting Up NetStumbler

After NetStumbler starts, you may want to set the options to maximize your wardriving experience.

Figure 9-1 shows data from an actual wardriving session, shot after the session. Looking at the window, you can see a left and a right pane. The status bar beneath the panes provides some valuable information. The message in the middle of the status bar tells you how many access points are active. To the right of that is the status information. You can find descriptions of the possible status messages in Table 9-1. The last piece of information on the far right tells you how many networks NetStumbler found. In our case, it found 461. The number before the slash tells you how many networks meet the criteria or filter that you selected from the left pane. If you are looking at the main screen and not filtering anything, then the first and second number are the same. Anytime you select anything from the left-hand pane, the first number will change. For example, when I select the Encryption Off under Filters, the number is 253 of 461, or about 55 percent of my neighbor's networks don't use encryption. (You can get a closer look at the two panes later in the chapter, after we talk about the setup options.)

Table 9-1	Status Message
Message	*Description*
Card not present	Wi-Fi card not detected. Make sure you have installed a wireless NIC.
A device attached to the system is not functioning	Problem working with the Wi-Fi card. Switch interface mode on the device menu.
Not scanning	Scanning is not enabled. Click the arrow or start from the File menu.
No APs active	Wi-Fi card is working, but not detecting any networks at the time.
x APs active	Wi-Fi card is working and detecting x number of networks.
GPS: Acquiring	NetStumbler is receiving a message from the GPS.
GPS: Disabled	The GPS is disabled. Start it to record network coordinates.
GPS: Disconnected	The GPS was working but stopped. Check the GPS power.

Message	Description
GPS: Listening	NetStumbler is attempting to make a connection to the GPS.
GPS: No position fix	The GPS is working but cannot find a signal. Move the GPS or your laptop.
GPS: Port unavailable	The communication port is locked by another program, such as Streets & Trips. Close the other program and try again.
GPS: Timed out	A connection could not be made to the GPS. Try a different port or turn the GPS on.
GPS: N:x W:y	Indicates your GPS is working, and these are your coordinates.
x/y	Currently displaying the *x* AP in the list of *y* APs.

You'll also see that there is the usual Windows drop-down menus, such as File and Edit. There also are some icons that we will discuss shortly. The logical place to start is with the menus. Under the File menu, you see New, Open, Close, and Save As. These features work similar to any other Windows-based program. There is a Merge feature that allows you to merge a previous scan with the current one. This allows you to merge all your scans into one scan. Another option on the File menu is Export. We cover exporting files in Summary, Text, or Wi-Scan format later in the chapter. You can use File⇨Enable to start the scan when you previously disabled it. (If Enable is not checked, then it is not enabled.) Alternatively, you can use Ctrl+B to enable a scan.

There are many choices under the View menu. First, you can decide whether you want to display the Toolbar at the top of the window or the Status Bar at the bottom of it. You use the Split option to size the two panes. Of course, you can select the bar running between the two panes and drag it either left or right. Select either Large Icons or Small Icons depending on your eyesight. Similarly, select either List or Details to change the amount of information displayed in the right pane. Zoom In/Zoom Out is sometimes grayed out, but you can use it with the Signal/Noise view to zoom in or out. Should you wish, you can use the Arrange Icons to view the icons By Name or let the program do it when you select Auto Arrange. Also, you can use the Line up Icons option to line up the icons in the right pane in List view. You can save the defaults by selecting Save Defaults. Network Stumbler displays information using the 8-point MS Sans Serif regular font style. Don't like this font? Then change it by selecting Font. The last selection under View is Options, which we detail later in the chapter.

Use the Device menu to select the device when you have more than one to use for scanning. In Figure 9-2 you can see the drop-down menu with multiple devices to select for your scan.

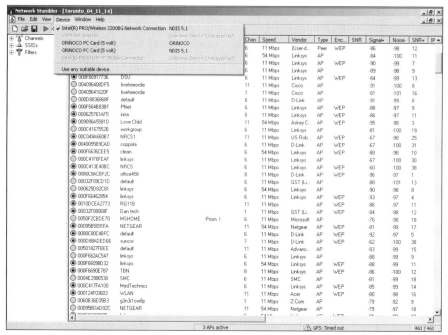

Figure 9-2:
Network
Stumbler
devices.

The Window menu allows you to adjust the window panes. And the Help menu offers the usual help information. (You may find older versions of NetStumbler that had no help information.)

Next, set up the options: Select Options from the View menu and you should see a dialog box like that shown in Figure 9-3.

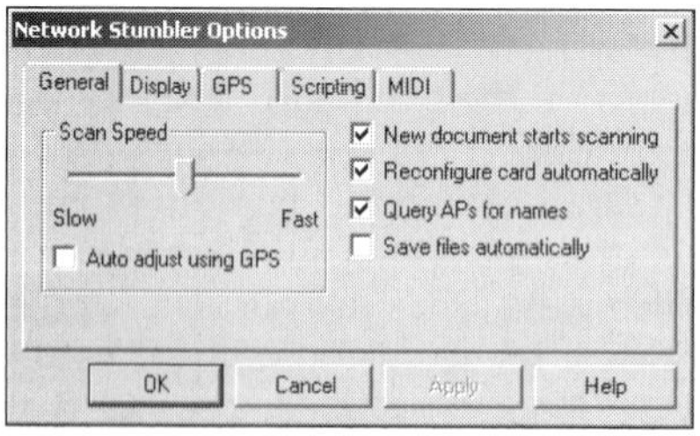

Figure 9-3:
Network
Stumbler
general
options.

The next subsections take a look at the tabs, starting with the General tab, which is shown on top when you select Options.

Selecting General options

For scan speed, there's a sliding scale from Slow to Fast on the left side of the General tab. Use the information in Table 9-2 as a rule of thumb for setting this parameter for your stumbling.

Table 9-2	Scan Speed Settings	
Setting	**Interval**	**Description**
Slow	1.50 seconds	For walking
-	1.25 seconds	For fast walking, jogging, and inline skating in a crowd
Midpoint	1.00 seconds	For inline skating and biking
-	0.75 seconds	For low-speed driving up to 25 Mph (about 40 Km/h)
Fast	0.50 seconds	For driving above 25 Mph (about 40 KM/h)

Table 9-3 describes the remaining parameters for the dialog box.

Table 9-3	General Scan Options
Option	**Description**
Auto adjust using GPS	Use this parameter to use your GPS positioning to determine the scan speed. It automatically adjusts the scan speed to the GPS velocity measurement. As your GPS reports speeds to NetStumbler, the timer frequency is set in the range of 2 to 6 times per second.
New document starts scanning	Use this parameter to force a new scan when you open a new document.
Reconfigure card automatically	Use this parameter to allow NetStumbler to reconfigure your wireless card using a null SSID and BSS mode. If you use this mode, then you may end up disassociated from an access point.
Query APs for names	Use this parameter to ask the device whether it supports names so NetStumbler can record the names.
Save files automatically	Use this file to save the current scan file automatically — every 5 minutes and when you close NetStumbler.

Those are the General options. Click the Display tab to see further options.

Selecting Display options

The Display options are really the Display *option,* since there is only one — the angle format. What you see in Figure 9-4 is a drop-down list controlling the GPS latitude-and-longitude format. The default value is degrees and minutes to the one-thousandth — in the format D°MM.MMM. The other options follow:

- ✔ Degrees to the ten-thousandth, in the format D.DDDD°

- ✔ Degrees to the hundred-thousandth, in the format D.DDDDD°

- ✔ Degrees and minutes, and ten-thousandths of a minute, in the format D°MM.MMMM

- ✔ Degrees, minutes, and seconds, in the format D°M_S_

- ✔ Degrees, minutes, seconds, and hundredths of a second in the format D°M_S.SS_

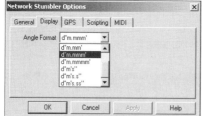

Figure 9-4:
Display
options.

You'll probably want to leave the default alone unless you have a compelling reason to change it (this assumes you understand why you may want to change it).

Selecting GPS options

Click the GPS tab and you should see a dialog box like the one in Figure 9-5.

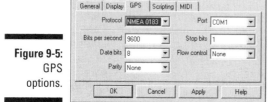

Figure 9-5:
GPS
options.

Table 9-4 lists the parameters, describes them, and provides the options or settings you may choose.

Table 9-4	GPS Options	
Option	*Description*	*Settings*
Protocol	Format of the GPS data	NMEA 0183, Earthmate, Garmin Binary, Garmin Text, or Tripmate
Bits per second	Transfer rate from the GPS	110 to 256000
Data bits	Number of bits used for data	5 to 8
Parity	Parity bits	Mark, One, Odd, or Space
Port	Communication port for the GPS	Disabled or COM1 to COM16
Stop bits	Number of bits used for communication	1, 1.5, or 2
Flow control	Handshaking protocol	None, Hardware, or Xon/Xoff

The NMEA standard sends a signal to NetStumbler every 2 seconds, whereas the Garmin standard sends it once per second.

Check the manual that comes with your GPS; it should tell you the settings you need.

Selecting Scripting options

NetStumbler lets you modify its operation through the use of scripts. You may choose to use common scripting languages such as PerlScript, Python, VBScript, Jscript, Windows Script Components, Windows Script Host, and Windows Script Runtime version. After you write your script, install it on the same system as Network Stumbler and then make it known by clicking the Scripting tab of the Network Stumbler Options dialog box. Do so and you should see the options shown in Figure 9-6.

Select the Type, File name, scripting Language, and Status of the script. Then when NetStumbler starts, it will execute the script. You can find a scripting guide at www.stumbler.net/scripting.html. Also, you might want to check out the Scripts Forum at

http://forums.netstumbler.com/forumdisplay.php?s=&forumid=24

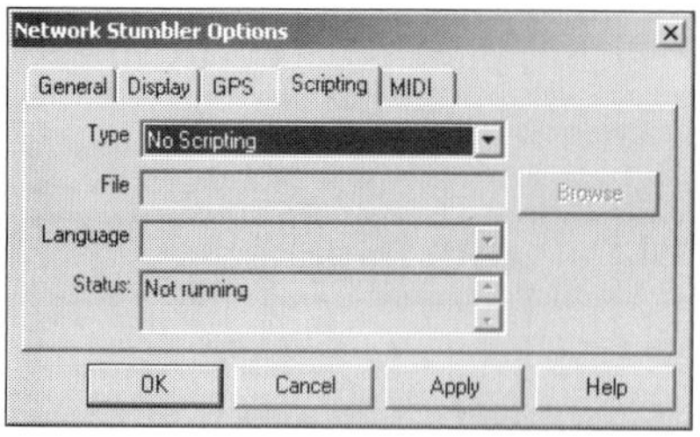

Figure 9-6:
Scripting
options.

Others have authored scripts and made them available through the Forum. For example, you can find a script to export NetStumbler output to Streets & Trips.

Selecting MIDI options

The final tab is for the MIDI or Musical Instrument Digital Interface settings. The MIDI standard is supported by most synthesizers. MIDI would allow NetStumbler to play music when events happen instead of the existing sounds. You could use this feature to modulate the sound as the signal gains or loses strength. Figure 9-7 shows the MIDI options.

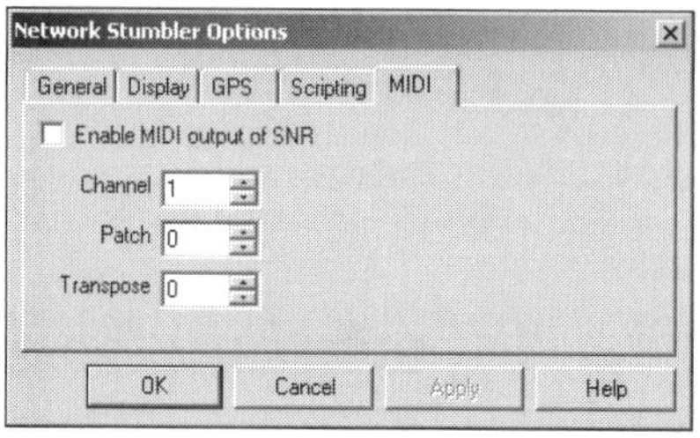

Figure 9-7:
MIDI
options.

First tick the Enable MIDI output of SNR box. Then you can change the MIDI Channel, Patch, and Transpose parameters. Check the manual that comes with your MIDI device for the correct settings for these parameters.

To change the existing sounds, you can also use your WAV files: Just rename them to the names used by Network Stumbler and move them to the Network Stumbler folder on your system.

Navigating the toolbar

Looking again at Figure 9-1, you can see some icons on the toolbar below the menu bar. Figure 9-8 shows you the icons from the toolbar.

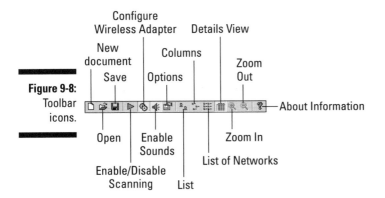

Figure 9-8:
Toolbar
icons.

The New (document icon), Open (folder icon), and Save (diskette icon) buttons are visible. You can use the green-arrow icon to enable or disable scanning. It works the same as selecting File⇨Enable. The gear icon automatically configures the wireless adapter. The hand-holding-the-menu icon opens the Options dialog box we talked about above. The two-underlined-documents icon enlarges the icon for the network shown in the right pane. It will also put them in columns as a list rather than one after another. The icon consisting of three small, underlined documents gives you smaller icons in columnar format. The six smaller underlined documents provide a list of the networks. The spreadsheet icon reverts the right-hand pane back to details view. You will find that the zoom in and out buttons are grayed out. The question mark provides About information. Click the X to close the About window should you open it.

If you need to change some of the options, you should do so now before we look at the results of our scan.

Interpreting the Results

NetStumbler provides a wealth of information, but it's just nonsense when you don't know how to interpret the data. So, okay, the first step in interpretation is to look back at Figure 9-1 and notice the two panes: The left pane is a familiar tree structure with three levels: Channels, SSIDs, and Filters; the right pane lists all detected networks. Table 9-5 lists the columns in the right pane and describes their usage.

Table 9-5	Right-Pane Column Headings and Descriptions
Column	**Description**
Circle Icon	You will notice a small circular or disk icon in the first column. When the icon has a padlock inside it, the access point uses encryption. Also, the icon changes color to denote signal strength. The color of the icon is one of the following: Grey — No signal Red — Poor signal Orange — Fair signal Yellow — Good signal Light green — Very good signal Bright green — Best signal
MAC	48-bit Media Access Code (MAC) address of the access point.
SSID	Network name or Service Set Identifier.
Name	The access point's name. This is an optional field, so frequently this field is blank. NetStumbler only detects the name of APs that use the ORiNOCO or Cisco naming standards.
Chan	Channel number the network is using. In North America, this number is between 1 and 11, though the standard specifies 1 through 14. An asterisk (*) following the channel number means NetStumbler is currently associated with the access point. When you see a plus sign (+), it means NetStumbler recently associated with the access point on the channel. When there is no character, it means NetStumbler located an access point but did not associate.
Speed	A misnomer for network capacity in Mbps (megabits per second). You will see either 11 (802.11 or 802.11b) or 54 Mbps (802.11a or 802.11g).
Vendor	Equipment manufacturer's name or other brand identifier.
Type	Network type, either AP or Peer. AP denotes an Infrastructure, Basic Service Set (BSS), or ESS (Extended Service Set) network. Peer denotes an Independent Basic Service Set (IBSS), Peer-to-Peer, or Ad-Hoc network.

Column	Description
Encryption	When the traffic is not transmitted in cleartext, you will see WEP in this column. NetStumbler cannot discern the type of encryption, but rather reports that WEP is on when the Flag is set to 0010 (the Privacy Flag).
SNR	The current Signal-to-Noise ratio, measured in microwatt decibels (dBm).
Signal+	The maximum RF signal seen.
Noise-	The minimum RF noise (the unusable part of a signal), shown in dBm.
SNR+	The maximum RF SNR in dBm.
IP Addr	The reported IP address of the device.
Subnet	The reported subnet.
Latitude	The latitude reported by the GPS when NetStumbler detects the network.
Longitude	The longitude reported by the GPS when NetStumbler detects the network.
First Seen	The time (based on the system's clock) when NetStumbler first detects the network, shown in hours, minutes, and seconds.
Last Seen	The time (based on the system's clock) when NetStumbler last detects the network, shown in hours, minutes, and seconds.
Signal	The current RF signal level, in dBm. You will see a value only when you are within range of a network.
Noise	The current RF noise level, in dBm. You will see a value only when you are within range of a network.
Flags	Flags from the network, in hexadecimal. Table 9-6 shows various values for the flags.
Beacon Interval	The interval of the beacon broadcast, measured in milliseconds.
Distance	The distance between where you currently are and the location when the best SNR was found. The default value is 100 ms, but you may see other values.

The first 24 bits (or 3 bytes) of the MAC or hardware address represent the manufacturer. The IEEE assigns these values, called Organizationally Unique Identifiers (OUI). You can find out an OUI for a manufacturer at

```
http://standards.ieee.org/regauth/oui/index.shtml
```

The displayed latitude-and-longitude values are actually *your* coordinates when you discover the network, not the actual coordinates of the network itself.

Table 9-6	NetStumbler Flags
Flag	**Description**
0001	BSS, ESS, or infrastructure mode.
0002	Peer-to-peer, IBSS, or ad-hoc mode. This is the inverse of the BSS mode.
0004	Connection Free (CF) polling for Request-To-Send/Clear-To-Send.
0008	Contention Free (CF) CF-Poll Request, used by the CF-Pollable protocol.
0010	Encryption is enabled.
0011	Infrastructure mode with encryption.
0020	WLAN uses the Short Preamble to improve the efficiency of some real-time applications such as streaming video or Voice over IP (VoIP).
0031	Infrastructure mode with encryption, using Short Preambles.
0040	WLAN uses Packet Binary Convolutional Code (PBCC). This indicates that the access point uses Texas Instruments' 22 Mbps version of 802.11b sometimes called 802.11b+.
0051	Infrastructure mode with encryption, using PBCC.
0080	Channel agility, which allows the network to switch channels automatically when there is interference.
0400	Short Time Slot.
2000	Direct Sequence Spread Spectrum (DSSS).
4000	Orthogonal Frequency-Division Multiplexing (OFDM).
DB00	Reserved for future use.

In the right pane, you can right-click a MAC address, and the Look Up options will show in a popup menu. If you can find an active network with an IP address or subnet value, this feature works; otherwise it won't. The Look Up options include a Look Up for ARIN (American Registry for Internet Numbers), RIPE (Réseaux Internet Protocol Européens), and APNIC (Asian Pacific Network Information Centre). Just select one of these to do a whois query on the address.

You can use the left pane to winnow down the data by channel, SSID, or the built-in filters. Clicking Channels aggregates the networks by channel numbers as shown in Figure 9-9. If you select channel 1, you can see that it displays the status of 35/461. Translation: Of the 461 networks, 35 used channel 1.

Similarly, we can click the + sign beside SSIDs and open it up to filter by network name. You can scroll the list. In Figure 9-10, we have highlighted SSID 101, the 3Com default. This shows us that 2 of the 461 networks use this SSID.

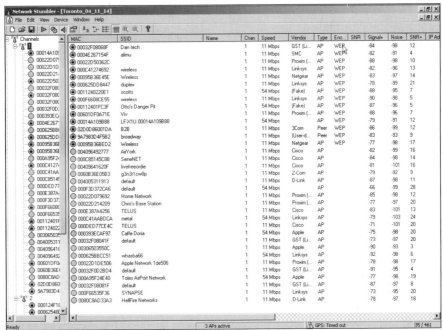

Figure 9-9:
Channel
display.

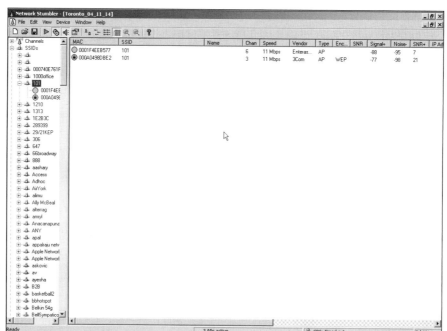

Figure 9-10:
SSID
display.

The last level in the left pane is Filters. If you look at Figure 9-11, you will see these nine built-in filters:

- ✓ **Encryption Off:** Shows only devices with WEP encryption disabled.
- ✓ **Encryption On:** Shows only devices with WEP encryption enabled.
- ✓ **ESS (AP):** Shows only devices in infrastructure mode.
- ✓ **IBSS (Peer):** Shows only devices in ad-hoc mode.
- ✓ **CF Pollable:** Shows only devices that are contention-free pollable.
- ✓ **Short Preamble:** Shows only devices with the Short Preamble enabled.
- ✓ **PBCC:** Shows only devices with PBCC enabled.
- ✓ **Short Slot Time (11g):** Shows only devices with a short slot time.
- ✓ **Default SSID:** Shows devices that are using the default SSID from the manufacturer.

If you don't know what these filters mean, then we recommend you get yourself a good introductory book on wireless networks. Peter recommends *Wireless Networks For Dummies* (Wiley).

Figure 9-11 shows the results sorted by the Encryption Off filter. In the figure, you can see that 253 of 461 networks have no encryption.

One last thing: Select Channel and then open one of the channels. Highlight a MAC address and you see a graphic representation of the Signal-to-Noise Ratio, as shown in Figure 9-12. The display shows red and green bars. The upper (or green) portion shows the RF signal above the noise, while the lower or red portion shows the noise level. Also, the decibels show as a negative number, measuring the power relative to one milliWatt (mW). You cannot see the purple line in the figure, but it's there — it indicates that the signal was momentarily lost because you moved out of range or something blocked the signal.

You can merge different NetStumbler files by choosing File⇨Merge and selecting the file(s) you want to merge with the current one. This way you can keep all your files together.

So there you have how to set up and use Network Stumbler. Now you can look at and study the information provided — but that's a lot easier to view when you plot the data on a map. As they say, "A picture is worth a thousand words."

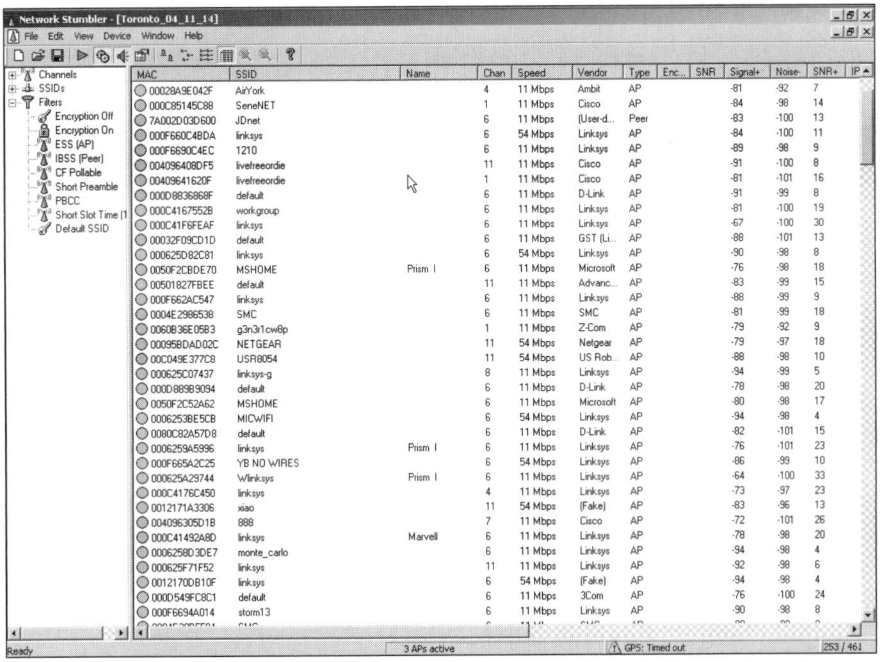

Figure 9-11:
Filters
display.

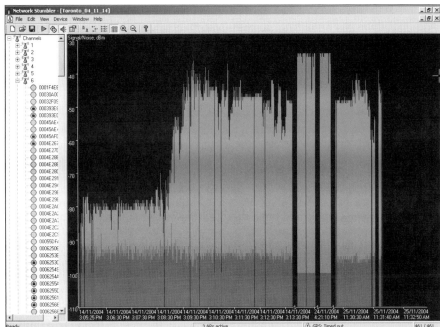

Figure 9-12:
Signal-to-
Noise Ratio
(SNR)
display.

Mapping Your Stumbling

So you finished your wardrive and you want to plot your data. Well, first you have to export it. This is as easy as selecting one of three options:

- ✔ **File➪Export➪Summary:** The Summary format exports the data in a tab-delimited format similar to that of the Network Stumbler graphical display. Choose Summary when you want to map the data in Microsoft's MapPoint and Streets & Trips.

- ✔ **File➪Export➪Text:** The Text format exports the same information but gives all readings for a particular network. Different signal strength readings create separate records. You might use this format to export the data to MySQL or Excel to do further analysis.

- ✔ **File➪Export➪wi-scan:** The Wi-Scan format exports the multiple readings for each network but with fewer columns. You can use the wi-scan format with Pete Shipley's Wi-Scan utility found at

      www.michiganwireless.org/tools/wi-scan/

Regardless of the format you choose, ensure that you append .txt to the filename. NetStumbler will not do it for you.

To use a map, you'll want to export the data using the Summary format. There are several ways to look at the data. In the following sections, we'll look at it using three different applications:

- ✔ StumbVerter and MapPoint
- ✔ Microsoft Streets & Trips
- ✔ DiGLE

Using StumbVerter and MapPoint

StumbVerter is a standalone freeware application you can use to import NetStumbler's Summary files into Microsoft's MapPoint 2004 maps.

Should you have an older version of MapPoint, you will need to download StumbVerter 1.0 Beta from

```
www.michiganwireless.org/tools/Stumbverter
```

Installing StumbVerter is as easy as installing any Windows program. Just run the `setup.exe` program and follow the steps to specify the destination folder and to verify the installation options. Figure 9-13 shows the opening window for StumbVerter.

To import the NetStumbler data you exported in the previous section, follow these steps:

1. **Click the Map icon and select *Create new North America* (or *Create new Europe*, whichever is appropriate).**

2. **Click the Import icon to open your Summary file and import it into StumbVerter.**

3. **From the Open window, highlight the exported file and click the Open button.**

StumbVerter will import your data and show the networks as small icons or pushpins, their colors and shapes relating to WEP mode and signal strength. You can download additional pushpins from `www.microsoft.com/downloads/details.aspx?familyid=2ad23c13-f367-45f4-809e-a77933eea57e&displaylang=en`.

MapPoint pushpins designate the access point, and by selecting one you can see balloons containing other information, such as the MAC address, signal strength, and mode. You can zoom in and out on the map.

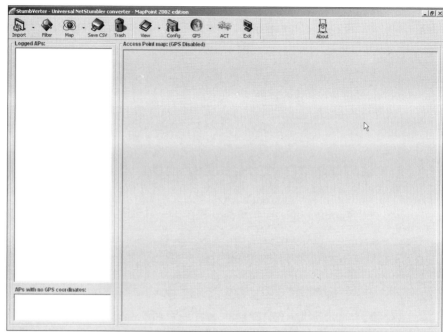

Figure 9-13:
StumbVerter
window.

When you are finished you can save the map as either a MapPoint (.ptm) document, HyperText Markup Language (HTML), or as a bitmap image. You'll need MapPoint to open a .ptm document, whereas you can open the HTML document by using Internet Explorer, Netscape Navigator, or Mozilla Firefox (or use Paint to open the open bitmap image).

You can find StumbVerter at `www.sonar-security.com/sv.html`.

Using Microsoft Streets & Trips

MapPoint is great but a little pricey. If you want to save a little money, you can use Microsoft Streets & Trips.

You have to perform an interim step before you can import your Summary file into Streets & Trips — parsing the Summary file. You could write a parser of your own or you can get a ready-made one at `http://kb3ipd.com/phpStumblerParser/index.php`. To use the phpStumblerParser, just click the Browse... button and navigate to the file on your system that you want to parse. Once you have selected the file, click the Generate Now! Button. Next, start Microsoft Streets & Trips.

To import the NetStumbler data you exported in the previous section, just select File⇨Open and navigate to the file you want to import. (This is the one you just parsed.) Figure 9-14 shows the Streets & Trips map for our wardrive.

Using DiGLE

Should you not want to line the coffers of Microsoft with your hard-earned cash, use DiGLE to generate your map. DiGLE stands for Delphi Imaging Geographic Lookup Engine. Go to the WiGLE registration page at `www.wigle.net/gps/gps/Register/main/`, fill it out, download DiGLE, and install it. Then double-click `digle.exe` to start the client shown in Figure 9-15.

Next you will need to get some maps by downloading a MapPack for the locale of your wardrive from

```
www.wigle.net/gps/gps/GPSDB/mappacks
```

Download the appropriate MapPack and unzip the contents into your DiGLE directory. There are map packs for every U.S. county and most major metropolitan areas.

After downloading a map pack, you're ready to import the NetStumbler data you exported in the previous section. Follow these steps:

1. **Use the First Choose drop-down list (shown in Figure 9-15) and select the map from the list.**

2. **Click the Load Local button.**

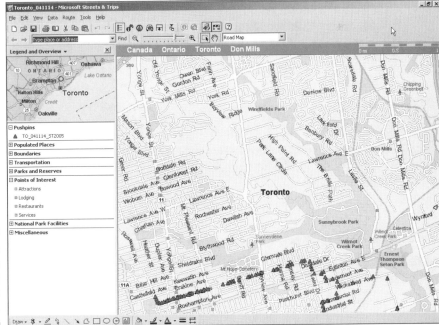

Figure 9-14:
Streets &
Trips
wardrive
map.

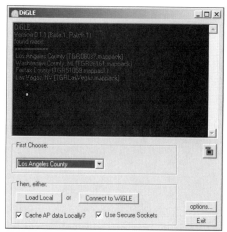

Figure 9-15:
DiGLE
window.

3. **Navigate to your Network Stumbler file, highlight the file, and then click the Open button.**

   DiGLE generates a map like the one shown in Figure 9-16.

   You can find DiGLE at `www.wigle.net/gps/gps/GPSDB/dl/`.

   If you don't use the Windows platform or want additional tools, you'll find the next chapter of interest.

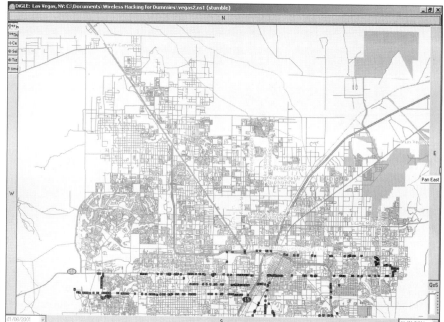

Figure 9-16:
DiGLE
mapping of
a wardrive.

Part III
Advanced Wi-Fi Hacks

The 5th Wave   By Rich Tennant

EXPERIMENTING WITH THE WIRELESS LASER BEAM NETWORK

"Okay—did you get that?"

In this part . . .

When you're off and running with your ethical wireless hacking tests, you can turn things up a notch. The tests up to this point in the book were more introductory in nature — less *technical,* if you will. In this part, we get into the nitty-gritty of wireless hacking.

For starters, we'll build on the wardriving techniques we introduced you to in Part II. Then we get into how to look for — and how to handle — unauthorized wireless devices on your network. We also take a look-see into common vulnerabilities — some that stem from the 802.11 protocol, and others that are network-based — all of which you can test for. Finally, we take a look at various denial-of-service attacks that can be carried out against wireless systems, and what you can do to protect against them.

Chapter 10

Still at War

Sun Tzu wrote *The Art of War* two and a half thousand years ago. It's a simple book, but a profound one that every security professional should study. Sun Tzu wrote: "If you know the enemy and know yourself, you need not fear the result of a hundred battles." In this chapter, we show you how to learn more about your network by showing you the information an outsider can easily obtain. Then it's up to you to do something. If you know yourself and your enemy, then you should not fear the result of anyone using the tools in this chapter.

Sometimes one tool is insufficient for your needs, and you need to supplement the tool with another. The tools in this chapter help you do a better job of network discovery.

Using Advanced Wardriving Software

We discuss how to use NetStumbler in the previous chapter. NetStumbler is a great tool — preferred by nine out of ten wardrivers — but it just doesn't give you everything you need. Additional applications like Kismet, Wellenreiter, and MiniStumbler provide features NetStumbler can't provide. For instance, NetStumbler does not tell you about "closed" systems or systems that don't broadcast their SSID, but Kismet does. NetStumbler is a simple beacon scanner, but Kismet is a *passive network scanner,* capable of detecting traffic from access points and clients. Also, you have to run NetStumbler on a laptop, portable, or luggable computer. But those devices are not *really* portable.

(Trust us — we spend a lot of time on the road lugging laptops around. They get *heavy.*) So we show you how to use MiniStumbler, which runs on a hand-held. The following sections give you all the details.

Installing and using Kismet

If you believe your destiny is to discover wireless networks, then Kismet is for you. Kismet is freeware 802.11b and g (and 802.11a with the right card) wardriving software. Kismet can capture data from multiple packet sources and can log in ethereal-, tcpdump-, and AirSnort-compatible log files. In addition, Kismet can do the following:

- ✔ Detect other scanning programs like NetStumbler
- ✔ Channel hop
- ✔ Highlight the detected default access point configurations
- ✔ Discover "closed," "hidden," or "cloaked" SSIDs for access points where SSID broadcast is disabled
- ✔ Identify the manufacturers of discovered access points
- ✔ Group and custom name SSIDs
- ✔ Detect Cisco products by using CDP
- ✔ Detect IP block
- ✔ Passively monitor and record wireless network data packets, including encrypted ones
- ✔ Map access point locations using a GPS
- ✔ Work with ethereal and AirSnort

Kismet runs on most UNIX-like systems, including Linux, Mac OS, and Cygwin, and supports Hermes and Prism2 chipset cards with linux-wlan-ng drivers. You can find information at the following Web sites:

- ✔ You can find more about drivers at Jean Tourrilhes' Web page:

   www.hpl.hp.com/personal/Jean_Tourrilhes/Linux/Wireless.html

- ✔ Mark Mathew's AbsoluteValue Systems Web page offers information about drivers as well.

   www.linux-wlan.com/linux-wlan

- ✔ If you feel adventurous, you can learn how to install Kismet on Cygwin:

   www.renderlab.net/projects/wardrive/wrt54g/kismetonwindows.html

Does my card support monitor mode?

You can determine whether your wireless interface supports monitor mode with your current drivers with one easy Linux command. Use the `iwpriv eth1` (or `wlan0` or whatever segment) command as root. This command shows you any potential driver options that your card loads when Linux boots. If you don't see monitor mode, you need to find and install the applicable driver patch.

You can find Kismet at www.kismetwireless.net. You also can get Kismet for handheld computers — that is, iPaq/ARM and Zaurus/ARM — with embedded Linux. You need the ARM version from www.kismetwireless.net/download.shtml.

Preparing to install Kismet

Before you install Kismet, you need to determine whether your wireless interface supports monitor mode. If it doesn't, you need to set it up so that it does; otherwise, you cannot use Kismet. Kismet even supports ar5k-based 802.11a cards.

If you have more than one wireless card, you can split the work of network scanning over the cards. The Kismet documentation provides information on this feature.

To get the most out of Kismet, you may want to make sure you have the following before getting started:

- ✔ **libpcap (www.tcpdump.org):** libpcap is a freeware program that facilitates the capturing and formatting of the frames. Kismet requires libpcap. Make sure you get a version that supports wireless sniffing.

- ✔ **ethereal (www.ethereal.org):** ethereal is the gold standard for Linux sniffing. It's not required, but is highly recommended that you use ethereal to analyze the capture files. We discuss ethereal in Chapter 8.

- ✔ **GpsDrive (www.kraftvoll.at/software):** GpsDrive is beggarware that provides GPS mapping. You can link Kismet to your GPS with this program.

Go ahead and download Kismet, and we'll explain how to install and run it and interpret the results.

Installing Kismet

The first step to installing Kismet is configuring it by using the `configure` script. Table 10-1 shows Kismet's configuration options. To adjust an option, append it to the ./configure command. The following command, for example, shows you how to use the first option:

```
./configure –disable-curses
```

This is the proper way to run this script — from the current directory (although you can specify the whole path to execute the path). In Table 10-1, you see that this command disables the curses user interface.

Table 10-1	Kismet Switches
Option Description	*Option Flag*
Disable the curses user interfaces	disable-curses
Disable ncurses panel extensions	disable-panel
Disable GPS support	disable-gps
Disable Linux netlink socket capture (Prism2/ORiNOCO patched)	disable-netlink
Disable Linux capture support	disable-wireless
Disable libpcap capture support	disable-pcap
Enable the system syspcap (not recommended)	enable-syspcap
Disable suid-root installation	disable-suid-root
Enable the use of WSP remote sensor	enable-wsp100
Enable some extra stuff for Zaurus	enable-zaurus
Force the use of local dumper code when ethereal is present	enable-local-dumper
Support ethereal wiretap for logs (substitute the path to ethereal for DIR)	with-ethereal=DIR
Disable support for ethereal wiretap	without-ethereal
Enable support for the Advanced Configuration and Power Interface (ACPI)*	enable-acpi

* You must have Advanced Configuration and Power Interface (ACPI) enabled in Linux for the enable-acpi *option to work.*

In Linux, people sometimes refer to an option as a *switch*. The terminology you use is your choice, but the latter is more commonly used.

When you have finished configuring Kismet with the script, you are ready to do the following steps:

1. **If you haven't already done so, log in as root.**

2. **From the command prompt, run `make dep` to generate the dependencies.**

3. **Run `make` to compile Kismet using the GNU C Compiler (gcc).**

4. **Run `make install` to install Kismet.**

Now you are ready to use Kismet . . . almost. You still need to install the free GPSD. GPSD is the Global Positioning System Daemon, which provides spatial information from a GPS. This is useful for wardriving, especially after-the-fact. Without a GPS, you'll have all these discovered networks but you won't know how to find them again. GPSD is available for download from Russ Nelson at `www.pygps.org/gpsd/downloads`. The following steps show you how to install GPSD:

1. **Download `gpsd-1.10.tar.gz` or the latest version from Russ's Web site.**

2. **Make sure you are root.**

   Do a `su -` if you're not root.

3. **Type `gunzip gpsd-1.10.tar.gz` to uncompress the downloaded file.**

4. **Type `tar -xvf gpsd-1.10.tar` to untar the file.**

5. **Change to the directory you just created by typing `cd gpsd-1.10`.**

6. **Type `./configure` to execute the configure script.**

7. **Configure the GPSD binaries by typing `make`.**

8. **Copy the binaries to where you want by typing `make install`.**

   You can make (no pun intended) sure that gps and gpsd are in the appropriate directories by issuing the `which gps` and `which gpsd` commands. The `which` output shows you the full path to the program so you can make sure you placed them appropriately.

9. **Turn off your computer and make sure your GPS is turned off, too.**

10. **Connect your GPS to your computer with the serial cable.** (Of course, you can use a USB GPS as well.)

11. **Turn on the GPS.**

   Give it time to acquire a signal.

12. **Reboot your computer.**

13. Start the GPS daemon by typing `gpsd -s 4800 -d localhost -r 2947 -p /dev/ttyS0.`

You need root privileges to start the GPS daemon.

This starts the daemon listening on port 2947. You can verify that it is running by port scanning, using the `netstat` or `ps` command, or typing `telnet localhost 2947`. Table 10-2 provides some gpsd command line options.

If you have a USB GPS, you should type `gpsd -p /dev/ttyUSB0`.

Table 10-2	gpsd Command Line Options
Option	*Description*
`-d`	Debug level; you must specify a level.
`-K`	Keep-alive flag.
`-p`	Full path for the serial or USB GPS device.
`-s`	Baud rate. The most common rate is 9600, but you can specify different rates. But do so only when you know your GPS supports the rate.
`-S`	Port number where you want gpsd to open a listener. This is not the "listening port" for the gps itself, but for the GPS daemon or host.

Configuring Kismet

Now are you ready to use Kismet? Well, not quite. You must first edit the Kismet configuration file, `kismet.conf`. Unlike other Linux programs, you need to configure Kismet before you use it. To configure Kismet, open and customize the `/usr/local/etc/kismet.conf` file using your favorite editor, for example, `vi`, `pico`, or `emacs`.

You need root privileges to edit the `kismet.conf` file.

You need to change at least the following options:

✔ **suiduser:** Look for the comment `# User to setid to (should be your normal user)`. As it says, type the name of a normal account, not root.

✔ **Support for your wireless card:** By default, Kismet is configured to support Cisco cards. If you don't have a Cisco card, you need to comment out the Cisco card support and then add your card. If you have an ORiNOCO or other Hermes chipset card, then uncomment the ORiNOCO line. Similarly, if you have a Prism2 card, then uncomment the Prism2 line.

✔ **Source:** By default, the ORiNOCO card uses `eth0` as the capture device, and the Prism2 card uses `wlan0`. If your system uses something else, like `eth1` or `eth2` for the ORiNOCO capture device or `wlan1` or `wlan2` for the Prism2, then you need to change the device in the configuration file. The format for this variable is *driver, device, description*. So, if you are one of those right-brained individuals, you might use `source=orinoco,eth1, AirPort`. If you have money to burn and can afford the best, then you might use `source=cisco,eth0,ciscosource`.

✔ **Channel-hopping interval:** By default, the channel-hopping interval (`channelvelocity`) is set to `5`. This is the number of channels monitored every second. By increasing this value, you can monitor more channels per second. If you drive fairly fast, a lower value is better — with a high value, you will fly by and not get a good reading on the channel. If you want to monitor only one channel, then set `channelhop` to false, and it won't hop.

✔ **GPS support:** If you set up the GPS in the manner shown in this chapter, then the `gpshost` defaults are fine. Kismet is configured to use a serial device and listen on port 2947. If you don't want to use a GPS, then change the `gps` value to `false`.

If you want to change the sound Kismet plays when it finds a new access point, then change the `sound_alert` variable in the `kismet_ui.conf` file. Just type in the full path to the new `.wav` file you want to play. If you wish, you can change the user interface colors by altering the `kismet_ui.conf` file. You can use black, red, yellow, green, blue, magenta, cyan, or white, as well as bold (`hi-`). For instance, when you're worried about your battery status and you want the information to pop right out on the screen, change the `monitorcolor` variable to `hi-red` in the `kismet_ui.conf` file. This should make it more visible.

Should you wish, you can even use the Festival text-to-speech engine to report discovered networks. A little audio feedback is much safer when driving because you don't have to look at the laptop screen. You can find the Festival engine at `www.cstr.ed.ac.uk/projects/festival`.

Starting Kismet

This time we're not kidding, you really are ready to use Kismet. Obviously, Kismet is harder to set up and use than NetStumbler (see Chapter 9). But you are now ready. You'd think that setting the `suid` to an account other than root that is the account that you'd use. Wrong. If you logged on with that account, then `su` to root.

If you try to use an account other than root to run Kismet, you don't have permission to set the PID number file (`kismet_server.pid`) in the `/var/run` directory. You need to gain root privileges. Most of the time, you use `su -`. However, do a `su` and not a `su -`. If you do the latter, you cannot write the dump file. This is because when you use the latter, you change to root in the root environment. When you use the former, you gain root privileges, but maintain your normal user environment.

Now, to start Kismet, just type `kismet`. You can start Kismet with server (`kismet_server`) and client (`kismet_curses`) options. If you don't know what options to use, type `kismet -help`. We have listed the options in Table 10-3 for your convenience.

Table 10-3		Kismet Options
Flag	*Option Name*	*Description*
-a	allowed-hosts <hosts>	Comma-separated list of hosts allowed to connect
-c	capture-type <type>	Type of packet capture device; e.g., prism2, pcap
-d	dump-type <type>	Dumpfile type (wiretap)
-f	config-file <file>	Use alternative configuration file
-g	gps	GPS server; port or off
-h	help	The help file
-i	capture-interface <if>	Packet capture interface; e.g., eth0, eth1
-l	log-types <type>	Comma-separated list of types to log; e.g., dump, cisco, weak, network, gps
-m	max-packets <num>	Maximum number of packets before starting a new dump
-n	no-logging	No logging: process packets only
-p	port	TCP/IP server port for GUI connections
-q	quiet	Don't play sounds
-s	silent	Don't send any output to the console
-t	log-title <title>	Custom log file
-v	version	Kismet version

Understanding the Kismet user interface

Figure 10-1 shows the information you get when Kismet is running. You can
see that Kismet has three frames:

- ✔ Network List
- ✔ Info
- ✔ Status

These frames are described in the following paragraphs.

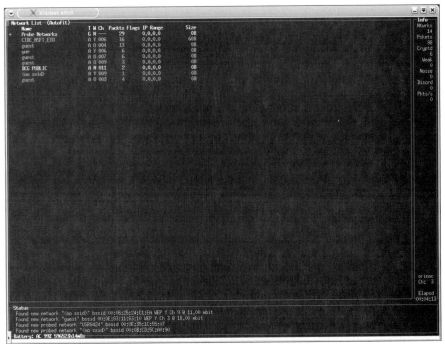

Figure 10-1:
Kismet
running.

Network List frame

In the Network List frame you see the fields described in Table 10-4. In its
help files, Kismet refers to this frame as the *Network display*. This frame takes
up the majority of the Kismet user interface. If you are using a GPS, you see
the GPS information in the bottom left-hand corner of this frame.

Table 10-4	Network Fields
Field	*Description*
Name	The BSSID or name of the wireless network sorted on the last time the network was seen
T	Type of WLAN detected: A = AP, H = Ad hoc, G = Group of wireless networks, D = Data only with no control packets, P = Probe request
W	WEP-enabled: Y = Yes, N = No
Ch	Channel of the device
Packts	Number of packets captured for the WLAN
Flags	Network attributes: A# = IP block found via ARP, U# = IP block found via UDP, D = IP block found via DHCP offer, C = Cisco equipment found, F = Vulnerable factory configuration
IP range	Self-explanatory
Size	Size of frame

If you see a ! (exclamation point) in front of a name, it's because Kismet saw activity in the last 3 seconds. If you see a . (period), that means Kismet saw activity within the last 6 seconds. Clearly, you are looking for those networks with an F flag!

Kismet starts in Autofit mode. In this mode, the names change automatically, and active ones appear first. However, you can sort the list by entering s at any time in the active window at the bottom. If you want to sort on SSID, enter ss. See Table 10-6 for the s command and others.

An interesting point on Kismet and encryption: Some access points don't accurately indicate the use of encryption by setting the correct bit in the 802.11 frame header. So, Kismet doesn't rely only on that bit. Instead, Kismet looks at the first few bytes of the logical link control (LLC) header to see whether they are the same. When they are, WEP is not used. When they are not, encryption is used.

Info frame

In the Info frame on the right-hand side, you see the fields described in Table 10-5. In the help documentation, Kismet refers to this frame as the *Statistics frame.*

Table 10-5	Info Fields
Field	*Description*
Ntwrks	Total number of WLANs detected
Pckets	Total number of packets captured
Cryptd	Total number of encrypted packets captured
Weak	Total number of packets with weak IVs (initialization vectors) captured
Noise	Total number of garbled packets captured
Discrd	Total number of packets discarded due to a bad CRC (ICV) value
Pkts/s	Number of packets captured per second
Ch:	Current channel
Elapsd (Discon)	Time (hours:minutes:seconds) elapsed since the start of the capture

You also see the type of card used for the capture. You will see `orinoc`, `prism`, or some other value.

Status frame

The status frame lists the major events detected by Kismet. It is a scrolling display. For example, you will see messages as it finds networks.

You may also see the battery status. This feature is helpful when you are walking around with your laptop and not plugged in.

Commanding Kismet

When you are running Kismet, you can use various commands to get more information. Table 10-6 provides the commands. If you're not sure of the commands, type h while Kismet is the active window.

Table 10-6	Kismet Commands
Command	*Description*
a	Show channel and encryption usage.
c	Show information about wireless clients associated with an access point.

(continued)

Table 10-6 *(continued)*

Command	Description
d	Print dumpable strings for quality, power, and noise.
e	List Kismet servers.
g	Group currently tagged networks together. Kismet prompts you for a new name.
h	Show the help window.
H	Return to normal channel hopping.
i	Get detailed information for a selected network.
l	Show wireless card power levels; that is, quality, power, and noise.
L	Stop channel hopping and stay on the current channel.
m	Mute sound.
n	Enter custom name.
Q	Quit.
r	Show the packet reception rate.
s	Sort network list.
t	Tag or untag current network or group.
u	Ungroup current group.
x	Close pop-up window.
z	Zoom network frame. This hides the info and status frames.

Kismet saves data automatically while it is running. When you are finished with Kismet, type Q to quit and close the application.

Kismet logging

By default, Kismet generates the following seven log files:

- ✔ **dump:** A raw packet dump.
- ✔ **network:** A plaintext log of detected networks.
- ✔ **csv:** A plaintext log of detected networks in comma-separated value format.

- ✔ **xml:** A plaintext log of detected networks in Extensible Markup Language (XML).

- ✔ **weak:** Weak packets detected and stored in AirSnort format. See Chapter 15 on what to do with this log.

- ✔ **cisco:** A log of Cisco equipment detected in Cisco Discovery Protocol (CDP) format.

- ✔ **gps:** A log of the GPS coordinates.

You can change logging by editing the `logtypes` variable in the `kismet.conf` file.

In Chapter 9, we discuss StumbVerter, which is an application that allows you to import NetStumbler's summary files into Microsoft's MapPoint 2004 maps. Well, StumbVerter doesn't work directly on Kismet logs. However, you can use a workaround: You can use WarGlue to convert your Kismet logs to NetStumbler format. From there you can export them to Summary format so StumbVerter can import them into MapPoint. You can get WarGlue at `www.lostboxen.net/warglue`. It's developed by the Church of Wi-Fi (CoWF). Or download WarKizniz from `www.personalwireless.org/tools/`. WarKizNiz also accepts input from Kismet log files and converts them into NetStumbler .ns1 format.

Shutting down Kismet

Because Kismet requires your wireless card to be in monitor mode, you can't use the card to connect to a wireless network without either restarting PC Card services or rebooting your system. Alternatively, you can run `kismet_unmonitor` as root and then eject the card and reinsert it. This should reset its original network parameters. If not, then restart.

Remember, you must put your wireless card into RF monitor mode to use Kismet so your card cannot associate with a wireless network. If you need a network connection while running Kismet, then either use another wireless card to connect to a wireless network or an Ethernet card to connect to a wired network.

Installing and using Wellenreiter

Wellenreiter is another wardriving tool. (A German-to-English dictionary translates *Wellenreiter* as *surfer*. Amazing how the subject of surfing keeps coming up in networking terminology.) Wellenreiter is a freeware user-friendly application, has a nice user interface, and is fairly straightforward to install and use.

Two versions of Wellenreiter are available:

✔ PERL scripts that support Linux. The script requires the Net::Pcap module (http://search.cpan.org/~kcarnut/Net-Pcap-0.05/) and the GTK or GIMP Toolkit (www.gtk.org). The latter is most likely already installed on most UNIX-based systems.

✔ C++ version that supports the wider UNIX world (C++/Qt) and handhelds such as Zaurus (C++/Opie).

The project is moving from PERL to C++.

Wellenreiter automatically handles the configuration and monitoring mode for most Cisco, ORiNOCO, and Prism2 cards. If you are using the PERL version, you can download, install, and run Wellenreiter by performing the following steps:

1. **Download Wellenreiter-v1.9.tar.gz or the latest version from the Wellenreiter Web site (www.wellenreiter.net).**

2. **Make sure you are root.**

   If you're not, do a su -.

3. **Uncompress the downloaded file by typing gunzip Wellenreiter-v1.9.tar.gz.**

4. **Untar the file by typing tar -xvf Wellenreiter-v1.9.tar.**

5. **Type cd Wellenreiter-v1.9 to change to the directory you just created.**

6. **Execute the Wellenreiter script by typing perl wellenreiter.pl.**

Surf on over to www.wellenreiter.net/screenshots.html to see some screenshots. Wellenreiter looks a lot like NetStumbler and Kismet, but we like its icons better. The left-hand pane lists the monitored channels, and the right-hand pane displays information about the wireless networks for the channels.

Wellenreiter saves a binary packet capture to the user's home directory. In our case — yours too, if you followed the steps above — the home directory is root, because we did the su to root. The binaries are in pcap format, which you can view with ethereal or tcpdump.

Using WarLinux

WarLinux is a Linux distribution for wardrivers. It's available on diskette and bootable CD. It is recommend for system administrators who want to audit

Wardriving, warwalking, and other war memes

The term *wardriving* was coined by Marius Milner as a play on the term *wardialing*. *Wardialing* in turn came from the 1983 movie *WarGames,* starring Matthew Broderick, Dabney Coleman, and Ally Sheedy. In the movie, Matthew, as nerdy David Lightman, unwittingly dials into a Department of Defense's war computer and almost starts a nuclear Armageddon. Forever after, hackers were portrayed as sitting at a computer connecting to networks.

Wardriving is also known as *NetStumbling* or *WiLDing* (*Wireless LAN Discovery*). (For more on WiLDing, see `www.bawug.org`.)

But what is warwalking? Well, wardriving is the meme for other forms of network discovery. Warwalking is one of the mutations. *Warwalking* (`http://wiki.personaltelco.net/index.cgi/WarWalking`) is network discovery by walking around. No longer are the hackers sitting at their computers. They're out and about in your neighborhood.

Here are some other terms you might hear about:

- *Warcycling* (`www.maths.tcd.ie/~dwmalone/p/sageie-02.pdf /`) is network discovery done from a motorcycle or bicycle.

- *Warflying* (`www4.tomshardware.com/column/20040430`) is network discovery done from an airplane. (Because many of the antennae are omnidirectional, you

actually get some very interesting information from the air.)

- *Warkayaking.* There have even been reports (`http://wifinetnews.com/archives/003922.html`) of warkayaking around Lake Union in Seattle, Washington.

- *Warchalking* (`http://forbes.jiwire.com/warchalking-introduction.htm` or `http://webword.com/moving/warchalking.html`) is the marking of the pavement to denote the existence of an access point. This variant seems inspired by hoboes who, using shared pictographs during the Great Depression, would denote easy marks and the active presence of railroad detectives in chalk. Warchalking, however, is for wibos, not winos.

- *Warspying* (`www.securityfocus.com/news/7931`) is when someone uses a X10 Wireless Technology receiver to capture the signals from wireless devices such as cameras. Makes you think twice about using those nanny-cams!

All you need to do is to think of a unique way to do network discovery to become famous. Hey, how about *warsurfing*? Not bad, but remember that water and electricity don't mix! (Oops. Actually, the term *warsurfing* [`www.netstumbler.org/showthread.php?t=2190`] was used to indicate the practice of using Google to find NS1 files on the Internet.)

and evaluate their wireless network installations. The benefit of WarLinux is that you don't have to install Linux but can boot it from a diskette or CD-ROM.

You can find WarLinux at `https://sourceforge.net/projects/warlinux`.

Installing and using MiniStumbler

NetStumbler, which we discuss in Chapter 9, was developed by Marius Milner. Well, Marius also developed MiniStumbler, a port for the Pocket PC. Marius calls NetStumbler and MiniStumbler "beggarware," his term for free-ware. MiniStumbler is commonly used when warwalking (see the sidebar, "Wardriving, warwalking, and other war memes") because it is versatile, user-friendly, and readily portable. It's versatile because you can run other programs like CENiffer to grab packets and discover networks. It is user-friendly because it uses a Windows user interface and is fairly easy to install and use. It is portable because you can carry it places you could not take your laptop.

MiniStumbler offers the same functionality as its big brother NetStumbler. Like NetStumbler, MiniStumbler sends probe requests every second. This is known as *active scanning,* contrasted to Kismet's passive scanning.

To use MiniStumbler, you need a handheld running HPC2000, Pocket PC 3.0, Pocket PC 2002, or Windows Mobile 2003. MiniStumbler supports Hermes (Avaya, Compaq, ORiNOCO, Proxim, TrueMobile) and Prism2 chipset cards, including Compact Flash. It also works with the built-in Wi-Fi of the Toshiba e740.

Due to its smaller screen size, MiniStumbler displays less information than its larger cousin. But the log files contain all the data that you get with NetStumbler. So, you can use MiniStumbler on your iPAQ to discover wireless networks, and use NetStumbler on your Windows XP laptop or desktop to view the data. Of course, you can dump the MiniStumbler output into StumbVerter (see Chapter 9) or use with Microsoft Streets & Trips.

MiniStumbler cannot read any of the text formats exported from NetStumbler.

Noticeably missing in MiniStumbler is MIDI support, so you cannot get that very gratifying audio feedback as it discovers access points.

You can find MiniStumbler at `www.netstumbler.com/downloads`.

Installing MiniStumbler

To install MiniStumbler, copy the proper processor architecture from your host computer to the Pocket PC. Supported processor architectures include ARM, MIPS, and SH3. Typically, the install goes as follows:

1. **Download MiniStumbler to the desktop.**

2. **To install the program from the desktop, run** `ministumbler installer.exe`**.**

   This action unpacks the files for the host computer and places the files in a special folder for the handheld.

3. **Select the directory where you want to install the program.**

   Use the default unless you have a compelling reason not to do so.

4. **Click Yes to see the Readme file. If you'd prefer not to view the file, click No.**

5. **Click OK when you see the reminder message to check the handheld device.**

6. **Connect your handheld to the host computer.**

   Activate your synching software if it doesn't start automatically. The software should upload the Pocket PC files to the handheld.

7. **If the files are in the `*.CAB` format, the Pocket PC where you uploaded the files will install the software.**

   There is no setup routine; you're set to go.

That should do it for you. However, you may need to review your documentation for full installation instructions.

Obviously, you need a handheld that supports a wireless adapter, whether it is a PC Card or a Compact Flash card. Installing the drivers for the wireless adapter usually requires the use of the synchronization program as well. You probably don't want to buy multiple wireless cards (even though we have), so you might have an ORiNOCO card you already use with Kismet and NetStumbler. Regrettably using an ORiNOCO card is one of the difficult installations. You have to download the driver for the card (use version 7.x or greater) and unpack the files manually. Then you have to manually port the files to the handheld and install the `*.CAB` file. Your device will warn you about unsafe drivers; just click OK. You did back up your Pocket PC, right? After you install the program and the drivers, you're set to go. Just locate the `ministumbler.exe` file under the Start or Start Programs menu and tap it.

Configuring MiniStumbler

After MiniStumbler executes, you should see the phrase No AP at the bottom of the window. This is good. Seeing No wireless is bad. (If you see the latter, make sure you have a working wireless card.) When MiniStumbler finds the first access point, you will see 1 AP. If your GPS is working, you will also see the message GPS on.

After you start MiniStumbler, two words and three icons appear on the menu bar. The two words are

- ✔ **File:** This menu holds the file functions such as Open and Save. Also, the Enable Scan item is on this menu. Use it to enable or disable network scanning.

- ✔ **View:** This menu holds those options that affect how the interface looks; for example, toolbars, icons, and options. We cover each of these in detail later in this chapter.

The three icons are

- ✔ **Green arrow:** Enables or disables scanning.
- ✔ **Gears:** Automatically reconfigures the wireless card for scanning. It stops the Wireless Zero Configuration service and makes sure the card is set to a blank or "ANY" SSID.
- ✔ **Hand with Menu:** Opens the Options screen.

The Hand with Menu icon gives you access to the following tabs:

- ✔ **General:** Find options about how the scan is performed. Make sure you check the Reconfigure card automatically and Get AP Names options. These options have the same meaning as in NetStumbler. You also find options for scanning speed. The speed ranges from Slower (1) to Faster (5). The default is Medium (3). Adjust it to 1. As a general rule, use the following guidelines for adjusting the speed:
  - • **Slower:** For warwalking.
  - • **Slow:** For jogging or leisurely inline skating.
  - • **Medium:** For inline skating or biking.
  - • **Fast:** For low-speed driving up to 25 mph.
  - • **Faster:** For driving over 25 mph.
- ✔ **Display:** This is a pull-down list box for display of GPS latitude and longitude. The default format is degrees and minutes to the one thousandth, that is, D°MM.MMM.
- ✔ **GPS:** In this tab, you find the information you need to make your GPS work with MiniStumbler. You can select the protocol, port (eight COM ports), baud, data bits, parity, stop bits, and flow control. The default is NMEA 0183, COM1, 4800 bps, 8 data bits, No parity bit, 1 stop bit.
- ✔ **Scripting:** You can write scripts to extend the functionality of MiniStumbler. Common scripting languages for Windows are VBScript and JScript.

If you have the correct configuration and a wireless device, MiniStumbler starts reporting as soon as you start the application.

Interpreting MiniStumbler

If you've used NetStumbler, you'll be comfortable with MiniStumbler. The user interface is essentially the same as the right-hand pane of NetStumbler. This means you have no filters to apply to the data; found in the left-hand pane of NetStumbler. MiniStumbler uses the same colors as NetStumbler. For example, detected networks show up as green, yellow, or red to indicate signal strength and gray for those out of range. The lock icon also means that the access point is WEP-enabled. Click the green arrow to pause discovery.

Unfortunately, MiniStumbler does not support the visualization tools that NetStumbler gives you. For example, you won't find the graph showing signal to noise over time.

Because of the small screen, you cannot see everything at once. You have to scroll to the right to see signal strength, SNR, and noise levels.

If you're set up to use a GPS, MiniStumbler also shows the latitude and longitude for all the wireless networks you find.

Selecting and right-clicking a Media Access Control (MAC) address opens a context menu that displays the SSID of the MAC. If you select an active MAC with an IP address or subnet assignment, then the menu provides three other Look up options regarding an IP address look-up at either ARIN, RIPE, or APNIC. Using this facility allows you to do a WHOIS query on the selected registration authority. It might help you determine whether you have a rogue access point.

You obviously need a network connection to the Internet to perform WHOIS queries on ARIN, RIPE, or APNIC.

When you exit MiniStumbler, it asks you whether to save the file. The file format is the same as that of NetStumbler: YYYYMMDDHHMMSS.ns1. Copy these files to NetStumbler and merge them with your other data.

Using other wardriving software

We know some of you out there for one reason or another don't want to use NetStumbler, MiniStumbler, Wellenreiter, Kismet, or WarLinux. One good reason is that your organization has a policy against the use of freeware or open source software. That alone would preclude the use of those programs. Other wardriving tools are available to you, however. Some are free and some have a fee. Here is a sample:

- Aerosol (www.sec33.com/sniph/aerosol.php)
- AirMagnet (www.airmagnet.com/products/index.htm)
- AiroPeek (www.wildpackets.com/products/airopeek)
- Airscanner (www.snapfiles.com/get/pocketpc/airscanner.html)
- AP Scanner (www.macupdate.com/info.php/id/5726)
- APsniff (www.monolith81.de/mirrors/index.php?path=apsniff/)
- BSD-Airtools (www.dachb0den.com/projects/bsd-airtools.html)
- dstumbler (www.dachb0den.com/projects/dstumbler.html)
- gWireless (http://gwifiapplet.sourceforge.net)

- iStumbler (`http://istumbler.net`)
- KisMAC (`www.binaervarianz.de/projekte/programmieren/kismac`)
- MacStumbler (`www.macstumbler.com`)
- Mognet (`www.l0t3k.net/tools/Wireless/Mognet-1.16.tar.gz`)
- NetChaser (`www.bitsnbolts.com`)
- Pocket Warrior (`www.pocketwarrior.org`)
- pocketWinc (`www.cirond.com/pocketwinc.php`)
- Sniff-em (`www.sniff-em.com`)
- Sniffer Wireless (`www.networkgeneral.com/`)
- THC-Scan (`www.thc.org/releases.php?q=scan`)
- THC-Wardrive (`www.thc.org/releases.php?q=wardrive`)
- WiStumbler (`www.gongon.com/persons/iseki/wistumbler/index.html`)
- Wireless Security Auditor (`www.research.ibm.com/gsal/wsa`)
- Wlandump (`www.guerrilla.net/gnet_linux_software.html`)

There is something for everyone on that list, regardless of whether you run Windows XP, Windows CE, SunOS, Red Hat Linux, FreeBSD, Mac OS, Zaurus, or Pocket PC.

Organization Wardriving Countermeasures

In this and the previous chapters, you find out how to use several wardriving programs to discover wireless networks. You're probably thinking you can't do much to protect yourself against these applications. Well, you can, and that is what the remainder of the chapter will show you.

Using Kismet

How can Kismet help with wardriving? Isn't it one of the problems? Don't people use it to discover my network? Well, yes, it is one of the problems. But look back at the description of Kismet. We said that you could use Kismet to find out whether others were running NetStumbler. Kismet sees and records the Probe Request. So here is your first countermeasure. Get yourself Kismet and look for others probing your wireless network. Commercial products like AiroPeek can help as well.

Disabling probe responses

When a workstation starts, it listens for beacon messages to find an access point in range to send a beacon to. Even though the access point is sending about 10 beacons a second, this is not always enough to detect them because the workstation has to monitor 11 channels by going to each channel and waiting 0.1 seconds before moving to the next channel. Further, when your authenticated access point's signal starts to weaken, your workstation needs to find another access point. For this reason, the 802.11 standard authors created the Probe Request. Your workstation can send a Probe Request, which any access point in range will respond to with a Probe Response. The workstation quickly learns about all access points in range. Now imagine that you're not trying to associate, but you're just trying to find access points in range, and you understand how NetStumbler works. So, the countermeasure is quite obvious: Turn off Probe Response on your access point.

Increasing beacon broadcast intervals

The beacon interval is a fixed field in the management frame. You can adjust the field to foil the fumbling stumblers. If someone drives by your access point and his or her device has to search all the channels and land on each for 0.1 seconds, then again the countermeasure is intuitive: Increase the beacon broadcast interval. This increases the likelihood that they won't grab your beacon when driving by.

Fake 'em out with a honeypot

The term *honeypot* harkens back to childhood: Winnie the Pooh and his love for honey. Perhaps you remember how he found a pot of honey, put his head in, and got stuck. Imagine this same concept applied to your wireless network. You put an attractive system on the network to draw hackers like a Pooh-bear to honey. Invite the hackers in. While the hackers are exploring the system, you watch them and try to learn about them or their behavior. You can learn about honeypots at `http://project.honeynet.org`.

Human nature suggests you might want to strike back when you find someone attempting to breach your security. This is not a good idea. You cannot fight back and you might not want to anyway. Crackers often take over other sites so you may harm an innocent party. If you have evidence that someone is attempting to break in, contact the Secret Service, the FBI, or your local law enforcement agency.

It's easy to set up a honeypot system. Install some access point software on a computer and then create directories with names like `Payroll` or `*wars`.

Turning the tables

As we often see, security tools are double-edged. Hackers have used Fake AP against hotspots. The hacker runs Fake AP on a laptop near a hotspot, say at a Starbucks. The clients wanting to use the Starbucks hotspot cannot discern the real access point from the cacophony of signals. This results in a denial of service to the hotspot's clients.

Don't turn on WEP and use a default SSID like `linksys`. A program like Fake AP (`www.blackalchemy.to/project/fakeap`) is useful for this purpose.

If one access point is good, then more is better. Black Alchemy developed Fake AP, which generates thousands of counterfeit 802.11b access points. Your real access point can hide in plain sight amongst the flood of fake beacon frames. As part of a honeypot or flying solo, Fake AP confuses NetStumblers and others. Because stumblers cannot easily determine the real access point, the theory is that they'll move on to the real low-hanging fruit — your neighbors. At least that is the theory. In real life, when you drive by a system with Fake AP, chances are it will not even register with NetStumbler. However, should you get stuck in traffic near the system, then that's a horse of a different color — you'll see the fake APs.

Fake AP runs on Linux and requires Perl 5.6 or later. You also need at least one Prism2 card with the CVS version of the Host AP Driver for Intersil Prism2/2.5/3 working. You can configure Fake AP to use dictionary lists for SSIDs and to generate WEP-encrypted and unencrypted access points.

If you're not Linux-inclined and prefer the Windows platform, you could use Honeyd-WIN32 (`www.securityprofiling.com/honeyd/honeyd.shtml`), which creates fake access points and simulates multiple operating systems. And if you have some change burning a hole in your pocket, try KF Sensor (`www.keyfocus.net/kfsensor/`).

Chapter 11

Unauthorized Wireless Devices

· ·

In This Chapter

▶ Understanding the consequences of unauthorized devices on your network

▶ Exploring basic wireless-network layouts

▶ Finding the common characteristics of unauthorized wireless devices

▶ Using various tools to search for unauthorized wireless devices

▶ Protecting against unauthorized wireless devices

· ·

A serious problem affecting wireless network security is the presence of unauthorized and rogue wireless devices. In Chapter 5, we discuss how employees and other users on your network sometimes introduce wireless equipment into your environment. These people unknowingly put your information at risk because they usually don't understand what can happen when they set up unauthorized access points and ad-hoc wireless clients. Even when users *do* understand the consequences, often they set up their own wireless networks anyway.

As a rule, people set up their own wireless systems because they want convenience — something that must be (but rarely is) balanced with security.

Most unauthorized wireless systems are not installed for malicious purposes. However, you must always be aware that people inside *and* outside your organization can introduce wireless devices for purely malicious reasons. Such unauthorized systems — commonly referred to as *rogue systems* — are often set up to gain access to your wireless data or cause other harm. The rogue wireless system of choice is an access point — both real APs and fake ones. These APs "lure" unsuspecting wireless-client systems to associate with them — and then pass all that wireless traffic through so everything can be captured and controlled elsewhere (usually for ill-gotten gain). We look at rogue devices in detail in Chapter 13.

In this chapter, we outline common characteristics of unauthorized wireless devices, various tests you can run to check for them, and what you can do to detect and prevent these systems in the future.

What Can Happen

There are serious vulnerabilities introduced if a person sets up unauthorized wireless systems on your network. There are four main types of potentially lethal unauthorized wireless devices threatening your airwaves:

- ✔ **Unauthorized APs:** Unauthorized APs are usually APs bought at a local consumer electronics store and installed onto your network without your knowledge. (See the scenario we outlined in Chapter 5.) This is the most common type of unauthorized system; when not properly secured, it can create a huge entry point into your network for *anyone within range*.

   A hacker can also use a Linux-based program called AirSnarf (`http://airsnarf.shmoo.com`) to create a legitimate-looking AP to which unsuspecting users can connect. The hacker can then attempt to grab usernames, passwords, and other sensitive information from the users, and they'll never know it happened.

   Many users are unaware of minimum wireless network security requirements in their organization, or they simply don't know how to properly configure such settings, which, in turn, create severe security problems.

- ✔ **Wireless clients:** These are usually laptops with wireless NICs that are set up to run in ad-hoc (peer-to-peer) mode. These systems don't require an AP to communicate with one another, and are often connected to the wired network — which (again) creates an easy backdoor into your system. These unauthorized systems can render other forms of security you have in place (such as firewalls, authentication systems, and so on) completely useless, exposing critical servers, databases, and other resources.

- ✔ **Rogue APs:** Rogue APs often are set up to mimic the characteristics of your legitimate APs in order to lure in unsuspecting users and wireless-client systems. Once these connections are made, the rogue AP can be configured to capture traffic and create denial-of-service (DoS) conditions. Chapter 14 covers DoS attacks and testing in more detail.

- ✔ **Rogue wireless clients:** These are unauthorized user systems (often external to your organization) that connect to your APs or ad-hoc systems and are usually up to no good.

Wireless System Configurations

Before we proceed, it makes sense to visually represent the various types of wireless network configurations we'll be searching for during our tests. The following list describes them in detail:

- ✔ **Basic Service Set (BSS):** This is the most common wireless network configuration. This setup, shown in Figure 11-1, consists of one AP and several wireless clients. In a BSS configuration, the AP serves as the network hub; all communications between clients go through it.

- ✔ **Extended Service Set (ESS):** This configuration (shown in Figure 11-2) includes multiple APs connected to the network, with roaming capabilities for mobile clients.

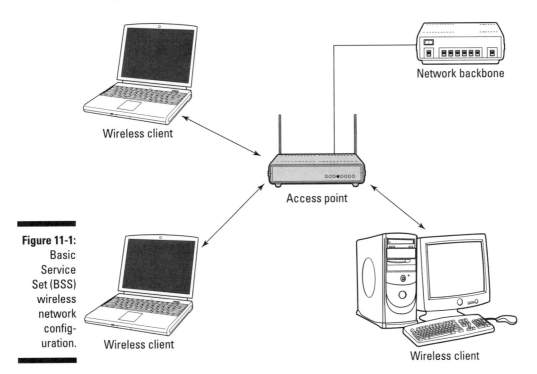

Network backbone

Wireless client

Access point

Figure 11-1:
Basic
Service
Set (BSS)
wireless
network
config-
uration.

Wireless client

Wireless client

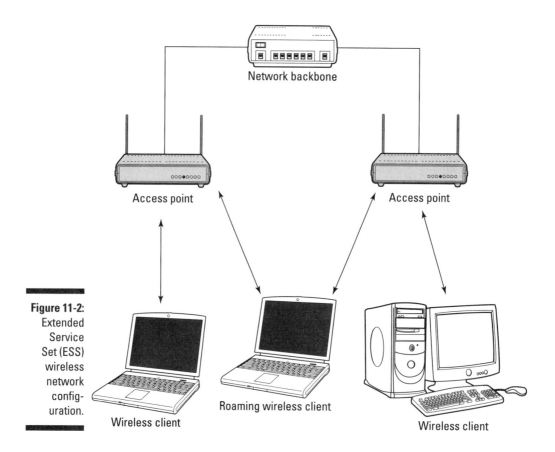

Figure 11-2:
Extended
Service
Set (ESS)
wireless
network
config-
uration.

Network backbone

Access point

Access point

Wireless client

Roaming wireless client

Wireless client

✔ **Independent Basic Service Set (IBSS):** This configuration is what we've been referring to as ad-hoc or peer-to-peer. This setup allows wireless clients to communicate to each other directly, without the need for a central AP to manage communications. An IBSS network is depicted in Figure 11-3.

It helps to know these wireless network configurations. Not only do they help focus your understanding of how each device communicates in a given network, they also give you a handle on the standard lingo that many wireless-network tools use when they refer to these systems.

Now, let's get down to business and start looking at common traits of unauthorized wireless devices.

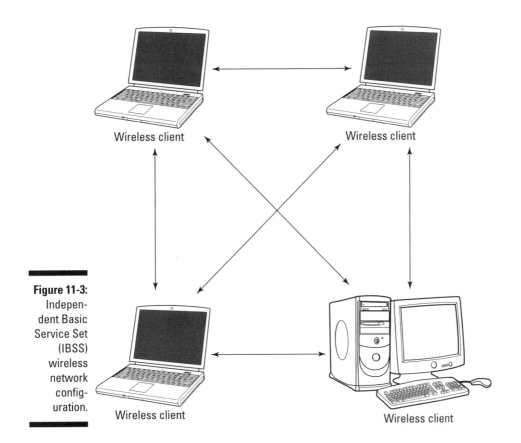

Figure 11-3:
Independent Basic Service Set (IBSS) wireless network configuration.

Wireless client

Wireless client

Wireless client

Wireless client

Characteristics of Unauthorized Systems

As we outlined in Chapters 9 and 10, it's pretty simple to perform a basic scan for wireless systems to see what's present on your network. However, it can be easy to overlook characteristics that point to unauthorized systems, especially if you have a large amount of hosts to sort though.

As with warwalking and wardriving (covered in Chapters 9 and 10), it's important to have the proper equipment to ferret out unauthorized systems. This includes a wireless NIC that supports all three 802.11 wireless standards — a, b, and g — as well as a good antenna that's sensitive enough to detect devices with weak signals.

During your quest for wireless devices that don't belong on your network, there are several common characteristics and issues we've found that can lead to unauthorized devices. Here are several items to keep in mind as you're performing your assessment:

✔ **Beacon packets where ESS field ≠ 1:** In 802.11 beacon packets, the last bit of the Beacon Capability Info field dictates whether the system is an IBSS or not. A zero (0) indicates an IBSS system (that is, it indicates a non-ESS type network) and potentially unauthorized system on your network. In addition, the next-to-last bit in the Capability Info field indicates the type of network. A one (1) indicates an IBSS system. These fields are shown in Figure 11-4.

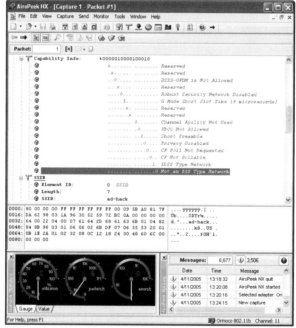

Figure 11-4: Beacon packet information indicating an IBSS (ad-hoc) system.

✔ Look for default SSIDs such as these:

   • `default` (common in D-Link APs)

   • `tsunami` (common in Cisco APs)

   • `comcomcom` (common 3COM APs)

   • `wireless` (common in Linksys APs)

   • `intel`, `linksys`, and so on (need we say more?)

Such SSIDs could indicate unauthorized systems on your network, especially if you're using a specific SSID that makes these odd ones really stand out.

✔ Also look for odd or strange-looking SSIDs, such as these:

- LarsWorld
- boardroom
- CartmansCubicle
- monkeybusiness
- HakAttak
- reception

✔ Unauthorized vendor hardware, especially those that show up as User-defined or Fake (as shown in Figure 11-5).

Figure 11-5:
NetStumbler
capture,
showing
what
appears to
be unautho-
rized vendor
hardware.

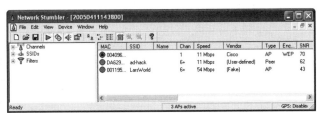

✔ Be on the lookout for MAC addresses that don't belong.

✔ You may also encounter network and protocol issues such as these:

- Odd or unsupported protocols (such as those for non-WEP traffic)
- Systems with consistently weak radio signals (low signal-to-noise ratios)
- Excessive numbers of packets transmitted at slower speeds
- Excessive DHCP requests or broadcasts
- Wireless network transmissions occurring during off-hours
- Excessive transmission retries
- Communications on different wireless channels
- Excessive CRC errors

We cover network and protocol issues like these in Chapters 12 and 13.

Although these wireless device characteristics are not a guarantee that you've got unauthorized systems on your network, they can be a good indicator and proof that you need to probe further. Keep in mind that just because you find what appear to be unauthorized wireless systems on your network, you've still got to figure out if they're actually in your building. If your organization has a standalone facility or campus, with no other buildings around, odds are the devices are on your network. However, if you share a building with other organizations, there's always a chance that the wireless devices you find are someone else's — and purely legitimate. This helps emphasize why you need to know your network — what's allowed, who's on it, etc.

Searching for unauthorized systems is often a matter of timing and luck. You may find nothing during some walkthroughs and several unauthorized systems during others. If at first you don't find anything suspicious, keep checking: The unauthorized system could be temporarily powered off at the time of your search.

Before we get started on using wireless software to track down unauthorized systems on your network, we thought it'd be a good time to mention a neat hardware solution for doing the same thing. This device is the handheld (actually key-chain-sized) WIFI Signal Locator by Mobile Edge (`www.mobile edge.com`). It's designed to determine whether a wireless hot-spot is in your vicinity, but you can also use it to sniff out unauthorized systems in your building as well.

Wireless Client Software

In Chapter 5, we demonstrated how you can use the basic wireless-network software built in to Windows XP to search for wireless systems. However, this method limits the amount of information you can ferret out when you're performing an extensive scan. The next best thing to use is the wireless client management software — such as ORiNOCO's Client Manager, Netgear Smart Wizard, and so on — that comes with your wireless NIC.

Figure 11-6 shows ORiNOCO's Client Manager discovering an ad-hoc network that's utilizing channel 6 for communication.

You may find a similar unauthorized system on your network. Unless your security policy allows users to have ad-hoc wireless devices (it doesn't, right?), the first tipoff that trouble's afoot would be the fact that you've got an ad-hoc network running. Also, you may have all your wireless systems set up to utilize another channel by default (such as channel 1) — so communications on channel 6 could indicate that this ad-hoc system is unauthorized.

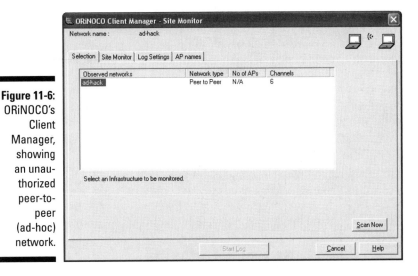

Figure 11-6:
ORiNOCO's
Client
Manager,
showing
an unau-
thorized
peer-to-
peer
(ad-hoc)
network.

Figure 11-7 shows Client Manager discovering an AP with a weak signal. (The weak signal is indicated by the small yellow bar in the SNR column.)

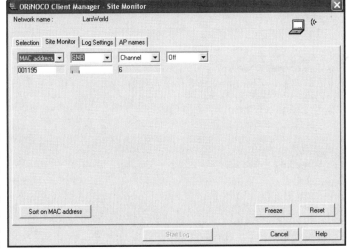

Figure 11-7:
Site
Manager,
showing an
AP with a
weak signal.

A weak signal can also indicate that you've got an authorized system that's far away or (cue the sinister music) that someone has turned the signal down on it and is trying to keep it hidden. You can walk around your office or campus using a utility such as Site Manager to see whether signal strength improves. If it does, you've likely narrowed down its location, so it's time to look into it further. If the signal doesn't improve, the AP may belong to someone else — but you still may have an unauthorized system on your hands.

As you can see in Figure 11-8, some client-manager software shows more detail than others. Notice how Netgear's Smart Wizard utility also shows signal strength, MAC address, *and* which 802.11 technology is being used — in this case, 802.11g.

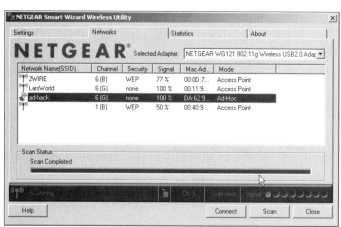

Figure 11-8:
Netgear's
Smart
Wizard
utility,
showing
an unau-
thorized
peer-to-
peer
(ad-hoc)
network.

Later in this chapter, we show you how you can use the MAC address of an ad-hoc system — along with a network analyzer — to track down specific IP addresses and protocols being used on the network.

Stumbling Software

The next step up, so to speak, in software you can use to detect unauthorized wireless devices is *stumbling software* such as NetStumbler and Kismet. Since we've already outlined how to use these programs in previous chapters, we'll spare you a repetition of those details. What's important to note here is the specific information you can find with a program such as NetStumbler.

Wireless network analyzers and monitoring tools such as AiroPeek and NetStumbler put your wireless NIC in *promiscuous monitoring mode* to capture all packets. This will effectively disable any other wireless communication (Internet, e-mail, network browsing, etc.) for that computer until you close out the program.

For starters, you can use NetStumbler to find unauthorized ad-hoc devices on your network. If you come across quite a few ad-hoc systems like the devices labeled `Peer` in Figure 11-9, you could be in for some trouble.

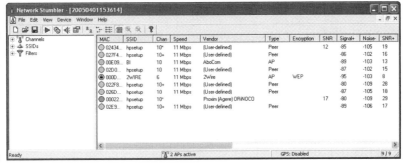

Figure 11-9: NetStumbler showing several unauthorized ad-hoc clients.

In the next section, we outline how you can use a network analyzer to determine whether the ad-hoc systems you find are attached directly to your network.

When using NetStumbler — or any wireless stumbling or analyzer software — the color of the indicator lights ranges from gray (no or minimal signal) all the way to green (strongest signal). This can tell you how close you are to the device in question.

In Figure 11-10, NetStumbler has found two potentially unauthorized APs. The ones that stand out are the two with SSIDs of `BI` and `LarsWorld`. Notice how they're running on two different channels, two different speeds, and are made by two different hardware vendors. Also, the ad-hoc system with vendor type "User-defined" looks suspicious as well. If you know what's supposed to be running on your wireless network, these devices really stand out as unauthorized.

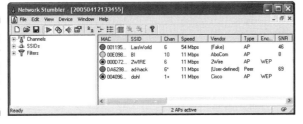

Figure 11-10: NetStumbler showing unauthorized APs.

You may remember from previous chapters that NetStumbler performs active probing of wireless systems. This means that if any APs are configured to disable beacon broadcasts — and thus disregard probe requests coming in from clients — then NetStumbler won't see them. Well, it'll see the AP along with its MAC address and associated radio information, but it won't see the SSID. Figure 11-11 shows what this looks like. Notice you can see everything about the Cisco AP but the SSID.

Figure 11-11:
A Cisco
AP with
a hidden
SSID in Net-
Stumbler.

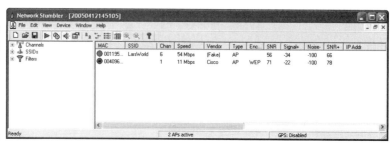

If you really need to see SSIDs that are "disabled", you can use the `essid_jack` hacking tool (outlined in Chapter 8) to create a client-to-AP re-association scenario that forces the SSID to be broadcast. Perhaps an easier way is simply to use a passive monitoring tool such as Kismet or a network analyzer.

Network-Analysis Software

Network analyzers, or *sniffers,* are great tools for seeking out rogue wireless equipment. Most network analyzers allow you to identify unauthorized systems — and can track down other information such as their IP addresses, what type of data they're transmitting, and more. Kevin's a little biased toward the ease-of-use offered by commercial analyzers, but many freeware and open-source tools will work just as well. Whatever your usability preferences, you can use network-analysis information to determine whether the systems you've detected are actually connected to your network — or if they're merely legitimate systems down the street or on the floor above.

Regardless of which network analyzer you use, you can still perform most of the basic functions we cover in this section.

Browsing the network

When seeking rogue wireless equipment with a network analyzer, we start out using AiroPeek NX to create what it calls a Peer Map. This map, shown in Figure 11-12, is essentially a physical layout of all wireless devices it can detect. When you know which wireless systems are out there talking, you're already on the trail of the rogues.

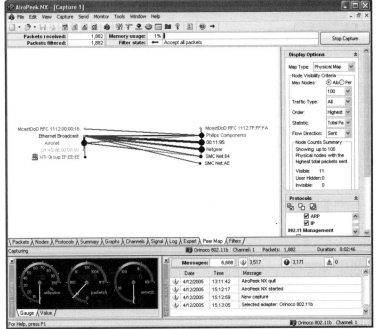

Figure 11-12:
AiroPeek
NX Peer
Map,
showing the
physical
layout of
surrounding
wireless
devices.

AiroPeek NX also has a feature it calls Expert analysis, which you can use to look for wireless anomalies (for example, ad-hoc clients and APs that don't belong). You simply load the program and select the Expert tab. Figure 11-13 shows the output of an Expert capture, along with its findings.

Notice that during this session, AiroPeek NX has found two rogue APs and 22 rogue ad-hoc clients! It can also find other helpful information such as APs that don't require WEP and those that are broadcasting their SSIDs.

AiroPeek and AiroPeek NX come with a security audit template called *Security Audit Template.ctf* that you can load to search for specific wireless security problems. This template can be loaded by simply clicking `File/New From Template`.

When performing a regular packet capture (such as the one shown in Figure 11-14), AiroPeek NX also points out wireless anomalies in the Expert column at the right. Notice that it found both ad-hoc clients *and* APs that don't belong. These functions can be helpful for pointing out the larger-scale security issues when you're scrolling through the seemingly overwhelming slew of packets you typically capture during a session.

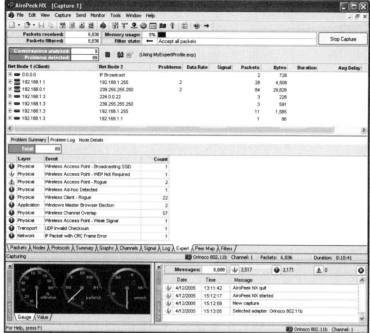

Figure 11-13:
AiroPeek
NX Expert
analysis,
showing
unau-
thorized
wireless
devices.

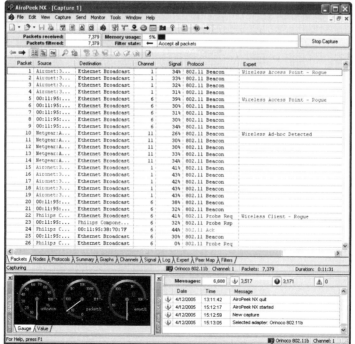

Figure 11-14:
AiroPeek
NX, point-
ing out
anomalies
during
a basic
packet
capture.

It's these types of features combined with general ease of use that make commercial tools such as AiroPeek NX and its sister application AiroPeek stand out. AiroPeek is discussed in greater detail in Chapter 8.

Probing further

In the previous sections, we outlined how to determine which wireless systems are transmitting radio signals in and around your organization. But how do you know if they're benign systems belonging to someone else outside your organization or are actually unauthorized systems connected to your network. There's one obvious way to find these systems — walk around and look for them. However, this may not be practical, especially if you have a large number of wireless devices or you're having trouble spotting them.

Let's look at how you can determine if an ad-hoc device is connected to your network. It's actually pretty simple by following these steps:

1. **Track down the MAC address of the system in question.**

   In this example, the system we want to check out is the one with the `Philips Components` address (as shown in Figure 11-15). We view this system by clicking the Nodes tab in AiroPeek NX. Note that AiroPeek NX displays the NIC vendor name in place of the first three bytes of the MAC address. (We've hidden the last three bytes just to provide our personal MACs some privacy.)

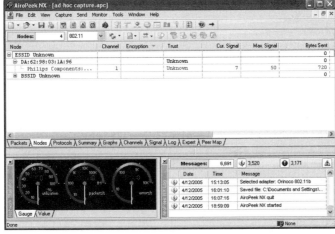

Figure 11-15: Using AiroPeek NX to find the MAC address of an ad-hoc system in question.

2. **Find the MAC address in the packets you've captured.**

   In AiroPeek NX, this simply involves switching to the Packets view by clicking the Packets tab and performing a hex search (by pressing Ctrl+F in AiroPeek NX) for the MAC address within the packets. In this example, dozens of packets were discovered; to keep things simple, we filtered out the unneeded management frames (beacons, probe requests, and so on) and focused on the IP-based traffic shown in Figure 11-16.

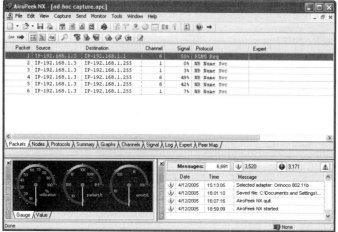

Figure 11-16:
Displaying
pertinent
IP-based
packets
in Airo-
Peek NX.

3. **Determine whether the associated IP address and protocols point to your network.**

   In this case, we found that the `Philips Components` MAC address has an IP address of 192.168.1.3. Now that you know the IP address, the next question is, *Is it a valid address on your network?* You may be surprised. Figure 11-16 shows this address, along with some interesting traffic — a PING Request and NB Name Svc broadcasts. This system is pinging another system (192.168.1.1 in this case) and appears to be a Windows-based computer — hence the NB (NetBIOS) broadcasts that tell the network *I'm here.* This type of traffic — especially if you know your users would never initiate it — could indicate an unauthorized system.

If you find a MAC address and you're not sure whether it belongs on your system, you can track down its IP address by matching it up to the IP-MAC address findings in SoftPerfect's Network Scanner (`www.softperfect.com/products/networkscanner`). This is a great way to match up MAC addresses to IP addresses and see if a system is on your network, and it's a lot quicker and simpler than performing reverse ARP lookups.

This test is not 100 percent foolproof, but it's a great test to run nonetheless. You can also use this method to determine whether unauthorized APs are connected to your network.

Additional Software Options

In addition to using the wireless-client, stumbling, and network-analysis software mentioned here, you have some additional ways to search for wireless devices that don't belong. For example, some basic port-scanning and vulnerability-assessment tools can give you useful results. Here's a quick list:

- ✔ SuperScan
- ✔ GFI LANguard Network Security Scanner
- ✔ Nessus
- ✔ NeWT
- ✔ QualysGuard

These programs aren't wireless-specific but they may be able to turn up wireless-device IP addresses and other vulnerabilities that you wouldn't have been able to discover otherwise.

Online Databases

One more place to look for unauthorized wireless systems is the Internet. (Well, yeah . . .) Up to this point, we've mentioned several Web sites you can browse to and query to see whether your "authorized" wireless devices have been made public — as in, plastered all over the Net. Well, you can also use these databases to search for unauthorized systems as well. If you know the exact GPS coordinates of your building, you can perform a detailed lookup in WiGLE's database at

```
www.wigle.net/gps/gps/GPSDB/query
```

to see whether any systems in your vicinity have been posted. If you don't mind sorting through entries by, city, state, or Zip code, you can also check out `www.wifimaps.com` and `www.wifinder.com` to see what you can find.

Unauthorized System Countermeasures

The countermeasures necessary to help prevent unauthorized wireless devices are similar to those we've discussed up to this point. They are:

✔ First and foremost, implement a reasonable and enforceable wireless security policy that forbids unauthorized wireless devices — and actually enforce it.

✔ Use stumbling software or a network analyzer to monitor for network changes and systems that don't belong.

✔ Use a full-fledged wireless intrusion-detection system (WIDS) or network-monitoring system that can find wireless network anomalies, prevent bad things from happening, and alert you in real time. Control access to authorized wireless devices only by one or more of the following:

- MAC address

- SSID

- Communications channel used

- Hardware vendor type

Chapter 12

Network Attacks

*Y*our computer systems and applications require one of the most funda-mental communications systems in your organization — your network. Although many organizations don't completely rely on wireless networks for everything, others do. Either way, your wireless network likely depends on critical servers; you can't afford to have them compromised via the network. These computers, even if they're an ancillary part of your overall network, are there for business reasons; damage them, damage the business. Therefore it's important to understand just what can happen when network-based 802.11 vulnerabilities are exploited.

There are thousands of possible network-level vulnerabilities on your wire-less systems — and seemingly just as many tools and testing techniques. The key point to remember here is that you don't need to test your wireless network for *every* possible vulnerability, using every tool available and tech-nique imaginable. Instead, look for vulnerabilities that can have a swift and immediate impact on your systems.

Some of the hacks and associated tests we demonstrate in this chapter are specific to 802.11. Others are security weaknesses common to any network — and those not only have a higher likelihood of being exploited, they can also have a high impact on your business.

No, it's not a Zip code

802.11 is a standard (in effect, a precise functional definition) that describes how a network can be accessed and controlled. The Institute of Electrical and Electronics Engineers (IEEE) establishes such standards and updates them, but so far no standard is perfect. Networks set up according to the 802.11 standard have certain characteristic weaknesses that bad-guy hackers use to get your network to give them (you guessed it) access and control. Ethical hackers must test and fix those vulnerabilities; this chapter describes them.

There are two main reasons that 802.11-based wireless systems are vulnerable at the network level:

- ✓ **Inherent trust allows wireless systems to come and go as they please on the network.** Practically everything about 802.11 is open by default — from authentication to cleartext communications to a dangerous lack of frame authentication. In addition to this equivalent of a "Hack Me" sign, wireless networks don't have the same layer of physical security present in wired networks.

- ✓ **Common network issues that 802.11 has inherited from its wired siblings enable attackers to exploit network-based vulnerabilities easily, regardless of the transmission medium.** The suspect activities allowed under 802.11 defaults include MAC-address spoofing, system scanning and enumeration, and packet sniffing. For openers.

Okay, some of the concepts in this chapter overlap material in other chapters in this book — and some of these vulnerabilities and tests could arguably be placed in other chapters that cover different categories of attacks. But our goal in this chapter is to give you the basis for a good overall assessment of your wireless systems at its most fundamental technical level — the network level.

What Can Happen

Network infrastructure vulnerabilities are the foundation for all technical security issues in your information systems. These lower-level vulnerabilities affect everything running on your network. That's why you need to test for them and eliminate them whenever possible.

Network-level attacks against wireless systems are usually simple to execute — but they have a high payoff. Though they may not be quite as disruptive as all-out denial-of-service (DoS) attacks, which we cover in Chapter 13, network-based attacks often lead to the compromise of wireless clients and APs — wreaking havoc on your business.

There's always the possibility that the tests we outline in this chapter can cause your wireless (and wired) networks to slow to a crawl — or crash altogether — so proceed with caution. If possible, perform your tests on non-production systems first — or perform them during times of non-peak network usage.

An important network test (covered in Chapter 7) is to see which systems are available and what security vulnerabilities are present on your wireless network. Previous chapters also harp (we think justly) on how important it is to keep your wireless network separate from your wired network. Unsecured wireless systems are about as safe as a screen door on a submarine.

When you know exactly what wireless systems are out there and where they're located, there are specific tests you can perform to exploit various vulnerabilities at the network level. This involves assessing such areas as MAC-address controls, whether or not a virtual private network (VPN) is in use, whether cleartext (unencrypted) communications are going on, which protocols are present, and more.

By exploiting these vulnerabilities, attackers can cause bad things to happen on your wireless network — these, for example:

✔ Attacking specific hosts by exploiting local vulnerabilities from across the network (which we cover in Chapter 7).

✔ Using a network analyzer to steal confidential information in e-mails and files being transferred.

✔ Gaining unauthorized access to your network.

Let's jump right into things and see these little nightmares in action.

MAC-Address Spoofing

A common attack carried out by hackers to circumvent basic access controls in wireless networks is to masquerade as a legitimate host on the network. They do it by *spoofing* (that is, faking and pretending to have) the identity of another system (which explains why this attack is sometimes referred to as a *wireless identity-theft attack*). Wireless NICs in clients, access points — basically any network device, wired or wireless — must have an identifier called a *MAC (media-access control) address*. This address is a 48-bit (six byte) number assigned by the component's manufacturer to make it unique. The idea is to identify the component (usually a specific network-interface card) to the host so switching, routing, and so on can happen without causing conflicts with other systems. No wonder someone who uses a fake address can make big trouble.

False sense of security

A popular — and pretty weak — security measure for wireless networks is to enable MAC address controls. This provides a form of AP authentication by allowing only clients with specific MAC addresses to access the wireless network. Sounds good, sure — and we often hear people saying, "I've enabled MAC address controls on my wireless network, so it's pretty secure." Well, actually, a hacker can circumvent this security measure very easily.

The IEEE calls the 48-bit MAC address space "MAC-48" — as originally published in the IEEE Ethernet specification. The first 24 bits (three bytes) of a MAC address make up a number unique to each NIC manufacturer. For example 00:40:5e belongs to Philips, 00:40:96 belongs to Aironet (now Cisco), and so on. Although this vendor identifier is called the Organizationally Unique Identifier (OUI), 16,777,216 OUIs are possible — and a vendor can have more than one. Each vendor can use the final 24 bits (three bytes) of the MAC address as desired, to create unique identifiers for all their cards (16,777,216 such identifiers are possible). The IEEE figures that all possible MAC addresses won't be exhausted any sooner than the year 2100.

You can look up the vendor ID of a specific MAC address at the following Web sites:

✔ http://standards.ieee.org/regauth/oui/index.shtml

✔ http://coffer.com/mac_find

Let's take a look at how MAC addresses can be changed on different platforms — and then we show you how a spoofing attack is carried out.

Changing your MAC in Linux

In Linux, you can spoof MAC addresses by following these steps:

1. **While logged in as** root, **disable the network interface so you can change the MAC address.**

   You do this by inserting the network-interface number that you want to disable (typically wlan0 or ath0) into the command, like this:

   ```
   [root@localhost root]# ifconfig wlan0 down
   ```

2. **Enter a command for the MAC address you want to use.** Here's how
   to insert the fake MAC address — and the network-interface number
   again — into the command:

```
[root@localhost root]# ifconfig wlan0 hw ether
        01:23:45:67:89:ab
```

The following command also works in Linux:

```
[root@localhost root]# ip link set wlan0 address
        01:23:45:67:89:ab
```

3. **Bring the interface back up with this command:**

```
[root@localhost root]# ifconfig wlan0 up
```

If you'll be changing your Linux MAC address(es) often, you can use a more
feature-rich utility called MAC Changer (`www.alobbs.com/macchanger`).

You can use the `ifconfig` utility in other flavors of UNIX as well. Refer to the
`ifconfig man` pages for specific parameters for your version.

Tweaking your Windows settings

If your test system is running Windows 2000 or XP, you may have several
options for changing your MAC address, depending on the wireless NIC you
have (and on its driver).

The first option to try is to see whether you can change the address by resetting your NIC's network properties. Here's the drill:

1. **Right-click My Network Places and then choose Properties.**

   A list of wireless NIC models appears.

2. **Right-click the maker and model of your wireless NIC and then choose
   Properties again.**

   The Properties window appears.

3. **Click Configure.**

   Another Properties window appears.

4. **Click the Advanced tab.**

   If your wireless NIC will allow you to change its MAC address, you'll
   have a Network Address (or MAC address) listed under Property.

5. **Click Network Address, click the radio button next to the Value field, and enter the 12-digit MAC address you want to use.**

Figure 12-1 illustrates this procedure.

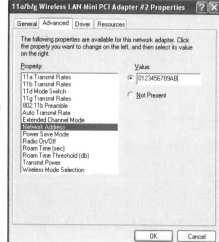

Figure 12-1:
Changing
the MAC
address in
a wireless
NIC's driver
settings in
Windows.

If your wireless NIC doesn't have the Network Address option, you can edit the Windows Registry to get the same result. Here's how to make it happen:

1. **At a Windows command prompt, enter** `ipconfig /all` **to view your current MAC address (as shown in Figure 12-2).**

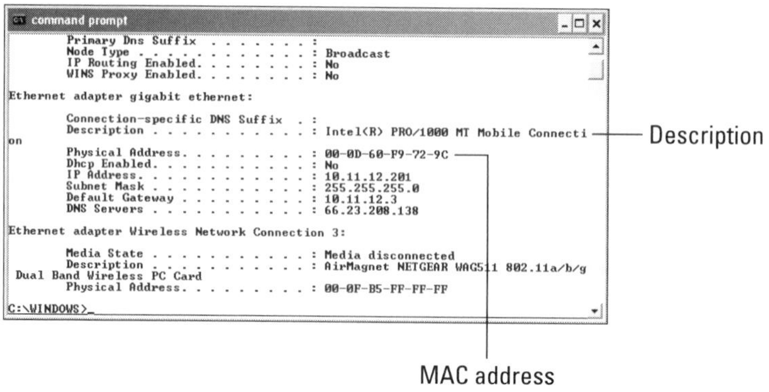

Figure 12-2:
Viewing
your cur-
rent MAC
address in
Windows.

2. **Run** `regedt32` **(not** `regedit`**) in Windows 2000, or** `regedit` **in Windows XP.**

   The Windows Registry opens, ready for editing.

3. **Make a backup copy of the Windows Registry as it is now.**

   This is a safety measure in case something goes awry and you have to restore it to its previous state.

   - If you're using `regedit` in XP, select File⇨Export.
   - If you're using `regedt32` in 2000, select Registry⇨Save Key.

4. **Browse to the key and expand it.**

   Here's the path to the key:

   ```
   HKEY_LOCAL_MACHINE\SYSTEM\CurrentControlSet\Control\Class\{4D36E972-
         E325-11CE-BFC1-08002BE10318}
   ```

5. **Find the subkey for the NIC you want to modify:**

   a. **Click through the various four-digit folders starting at 0000.**

      You're looking for the device that has a DriverDesc value that matches the Description shown when you enter `ipconfig /all` at a command prompt.

   b. **When you find the appropriate folder, expand it.**

6. **Right-click in the right window pane of the folder you've found, and then choose New⇨String Value.**

   A window appears, offering a place to enter a name for the folder.

7. **Enter** `NetworkAddress` **for the name.**

   Rest assured that a black-hat hacker would enter something more devious.

8. **Double-click the** `NetworkAddress` **key you just created, and then enter the new, 12-digit MAC address you'd like to use.**

   This new key is shown in Figure 12-3.

9. **Exit the Registry editor (**`regedit` **or** `regedit32`**).**

10. **Right-click My Network Places, click Properties, and choose Disable/Enable for the NIC you modified.**

    You can do this by simply right-clicking the NIC in the listing and selecting Disable, and then right-clicking again and selecting Enable. You can also reboot Windows — and may have to, depending on whether Disable/Enable works — to activate the new MAC address.

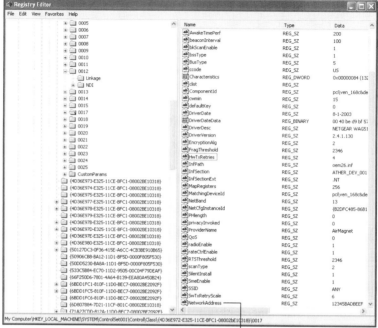

Figure 12-3:
Creating the
Network
Address
key in the
Windows
registry.

Network Address key

11. Verify that your change has taken place.

You do so by entering `ipconfig /all` at a Windows command prompt
again, as shown in Figure 12-4.

Figure 12-4:
Viewing
your new
MAC
address in
Windows.

Because MAC address changes are not immediate in Windows, you can use a
tool called DevCon by Microsoft — which is essentially a command-line version of the Device Manager utility for Windows 2000 and XP — to reset your

wireless NIC to make your Windows MAC-address changes immediate. DevCon is available for download at

```
http://support.microsoft.com/default.aspx?scid=kb;en-us;311272
```

SMAC'ing your address

If your wireless NIC driver doesn't allow MAC address changes as described in this chapter — or if you don't like editing the Windows Registry manually to change your MAC address — there's a neat and inexpensive tool you can use by KLC Consulting called SMAC (presumably short for *Spoof MAC*) at www.klcconsulting.net/smac.

Follow these steps to use SMAC:

1. **Load the program.**
2. **Select the adapter for which you want to change the MAC address.**
3. **Enter the new MAC address in the New Spoofed MAC Address fields and then click Update MAC.**
4. **Stop and restart the network card with these steps:**

   a. Right-click the network card in Network and Dialup Connections.

   b. Select Disable and then right-click again.

   c. Click Enable to put the change into effect.

   You may have to reboot for this to work properly.
5. **Click Refresh in the SMAC interface.**

   You should see a screen similar to the one shown in Figure 12-5.

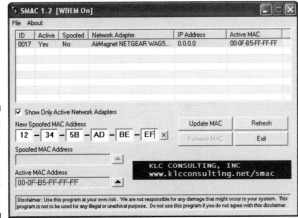

Figure 12-5:
SMAC showing a spoofed MAC address.

KLC Consulting also has a command-line version of SMAC that can be integrated with Microsoft's DevCon tool (mentioned earlier) for a complete solution to MAC address changes — and to resetting your hardware on the fly.

To reverse any of the MAC address changes shown here, simply reverse the steps performed and delete any data you created.

A walk down MAC-Spoofing Lane

So you've enabled MAC-address controls on your wireless network — but you're curious: Just how effective are those controls? Unfortunately, not very. Your wireless network is still vulnerable to unauthorized access, even though you've enabled MAC address filtering on your APs. Of course, if you don't have WEP, WPA, or some other form of encrypted communications in place, anyone with a wireless network analyzer (such as CommView for WiFi or AiroPeek) will still be able to view unencrypted traffic — all they have to do is jump through a couple more hoops, and they're in. By bypassing MAC address controls and obtaining an IP address, they can easily become part of the network. Once this occurs, an attacker can gain full access to your airwaves — and anything's fair game.

Come along with us, and we'll show you how you can test your MAC-address controls — and demonstrate just how easy they are to circumvent. Here's the procedure:

1. **Find an AP to attach to.**

   That's easy: Simply load NetStumbler, as shown in Figure 12-6.

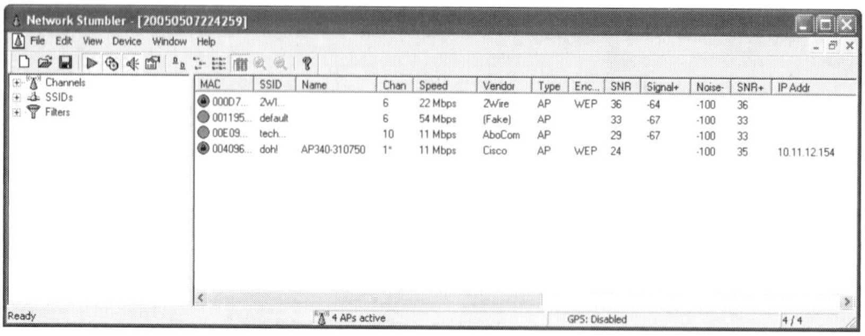

Figure 12-6:
Finding an accessible AP via Net-Stumbler.

You could skip Step 1 and just look for Probe Requests, but it's always good to make certain you're working with *your* wireless systems and not messing around with your neighbors' stuff. Instead of waiting to look for Probe Requests to get a valid MAC address, you could send out a Deauthentication frame to the broadcast address. This would force any wireless client within range to reauthenticate and reassociate to the AP revealing their MAC addresses in the process. You have to be careful doing this though so as not to disturb your neighbors' systems. We cover deauthentication and disassociation in Chapter 13.

In our "test" organization, shown in Figure 12-6, we know that the AP with an SSID of doh! is a valid one to test because that's the SSID we use on our network. Take note of the MAC address of this AP as well. Doing so helps you make sure you're looking at the right packets in the steps that follow. Although we've "hidden" most of the MAC address of this AP for the sake of privacy, let's just say that the MAC address you're looking for here is 00:40:96:FF:FF:FF. Also notice in Figure 12-6 that NetStumbler was able to determine the IP address of the AP. Getting an IP address helps us confirm that we're on the right wireless network.

One simple way to determine whether an AP has MAC-address controls enabled is to try to associate with it so you can obtain an IP address via DHCP. If you *can* get an IP address, then the AP doesn't have MAC-address controls enabled. Now, for security's sake and if you so desire, take a few minutes to go turn on MAC-address controls on your AP(s) — you can come back and run this test again to verify that you cannot obtain an address via DHCP.

2. **Using a wireless network analyzer, look for a wireless client sending a probe request packet to the broadcast address — or for the AP replying with a probe response.**

You can set up a filter in your analyzer to look for such frames, or simply capture packets and browse through them, looking for your AP's MAC address as noted earlier. Figure 12-7 shows what the Probe Request and Probe Response packets look like.

Note the wireless client (again, for privacy, let's say its full MAC is 00:09:5B:FF:FF:FF) first sends out a Probe Request to the broadcast address (FF:FF:FF:FF:FF:FF) in packet number 98. The AP with the MAC address we're looking for replies with a Probe Response to 00:09:5B:FF:FF:FF that confirms this is indeed a wireless client on the network for which we'll be testing MAC-address controls.

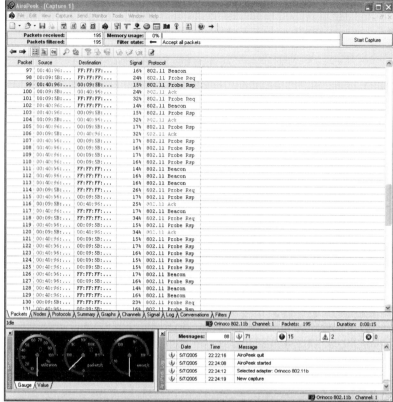

3. **Change your test computer's MAC address to that of the wireless client's MAC address (the one you found in Step 2 of these instructions).**

You can verify your new MAC address as shown by running `ipconfig /all` at a Windows command prompt, as shown in Figure 12-8.

```
C:\WINDOWS>ipconfig /all

Windows IP Configuration

        Host Name . . . . . . . . . . . . : PC1
        Primary Dns Suffix  . . . . . . . :
        Node Type . . . . . . . . . . . . : Broadcast
        IP Routing Enabled. . . . . . . . : No
        WINS Proxy Enabled. . . . . . . . : No

Ethernet adapter Netgear WAG511:

        Connection-specific DNS Suffix  . :
        Description . . . . . . . . . . . : AirMagnet NETGEAR WAG511 802.11a/b/g
Dual Band Wireless PC Card
        Physical Address. . . . . . . . . : 00-09-5B-FF-FF-FF
        Dhcp Enabled. . . . . . . . . . . : Yes
        Autoconfiguration Enabled . . . . : Yes
        IP Address. . . . . . . . . . . . : 0.0.0.0
        Subnet Mask . . . . . . . . . . . : 0.0.0.0
        Default Gateway . . . . . . . . . :
        DHCP Server . . . . . . . . . . . : 255.255.255.255

C:\WINDOWS>_
```

Note that APs, routers, switches, and the like should be able to detect when more than one system is using the same MAC address on the network (yours and the client that you're spoofing). You may have to wait until that other system is no longer on the network or send a Deauthenticate packet to knock it off as shown in Chapter 13. However, we've seen very few quirky issues emerge from spoofing a MAC address in this way, so you may not have to do anything at all — it's likely to work without any problems.

4. **Ensure your wireless NIC is configured for the appropriate SSID.** For this example, we'll set the SSID to doh! (as shown in the Netgear Smart Wizard utility in Figure 12-9).

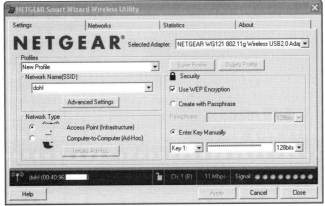

Figure 12-9: Ensuring that your SSID is correctly set.

Even if your network is running WEP, as is the case here, you can still test your MAC address controls. You'll just need to enter your WEP key(s) before you can connect.

5. **Obtain an IP address on the network.**

You can do this by rebooting, or disabling/enabling your wireless NIC. However, you can do it manually as shown by running ipconfig /renew.

Because we know the IP addressing scheme of the wireless network in this example (10.11.12.*x*), we could also manually set our IP address and get on the network.

6. **Confirm that you're on the network by pinging another host or browsing the Internet.**

You can do this by pinging the AP (10.11.12.154) or by simply loading your favorite Web browser and browsing to your favorite site.

That's all there is to it! You've circumvented your wireless network's MAC-address controls in six simple steps. We told you it was easy.

Who's that Man in the Middle?

Man-in-the-middle attacks — referred to as MITM or *monkey-in-the-middle* attacks (taken from a popular MITM tool called `monkey_jack`) — are network-level attacks whereby the attacker (the monkey) inserts his system in between a wireless client and an AP, as shown in Figure 12-10.

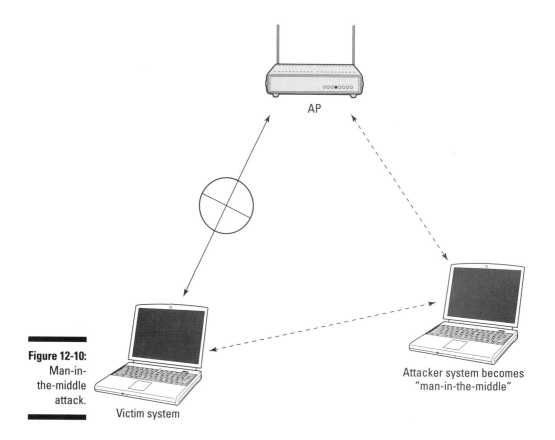

AP

Attacker system becomes "man-in-the-middle"

Figure 12-10: Man-in-the-middle attack.

Victim system

These attacks are slightly more theoretical (less practical) and definitely more difficult to carry out than other network attacks. However, once an attacker has inserted himself as the man-in-the-middle, he can do it again — and do various unpleasant things, including

- ✔ Capture data
- ✔ Inject new packets into the data stream
- ✔ Manipulate encryption mechanisms in IPsec, SSL, SSH, and so on

✔ Delay wireless communications

✔ Deny wireless communications

✔ Redirect traffic to a malicious application

The attacker can exploit MITM vulnerabilities in standard unencrypted wireless sessions as well as 802.1x EAP and PEAP sessions. It's even possible for an attacker to perform MITM attacks that exploit management packets — even when the wireless victims are running WEP or WPA.

Wireless hackers can exploit MITM vulnerabilities regardless of whether the communication is encrypted.

These attacks can happen in various ways such as

✔ **ARP poisoning:** This manipulates OS, router, and switch ARP tables so an attacker can spoof a victim's MAC address.

✔ **Port stealing:** Here an attacker can spoof packets by setting the source address to his victim's address and the destination address to his own address. In effect, the hacker takes control of his victim's traffic.

There are various tools that hackers can use to create MITM attacks. The most popular MITM tools are open-source tools for the UNIX/Linux and Windows platforms (in the case of Ettercap).

✔ Airjack suite (`http://sourceforge.net/projects/airjack`), which includes monkey_jack for automated wireless MITM attacks.

✔ dsniff (`www.monkey.org/~dugsong/dsniff`)

✔ Arpmim (`http://packetstorm.linuxsecurity.com/groups/teso/arpmim-0.2.tar.gz`)

✔ Ettercap (`http://sourceforge.net/projects/ettercap/`)

You can, of course, use these same utilities to test your wireless systems in an ethical-hacking fashion — but again, be careful.

Performing MITM attacks against your wireless network can be hazardous to your network's health. If one of those goes awry, it can redirect traffic, disconnect clients, and even create denial-of-service conditions. Proceed with caution.

Management-frame attacks

The first type of wireless MITM attack is an attack against various 802.11 management frames. As we've discussed in other chapters, 802.11 specifies no inherent authentication of management frames — and MAC addresses are simple to spoof — which makes this a popular wireless attack.

A MITM attack that exploits 802.11 management-frame vulnerabilities can be executed via the following steps:

1. The attacker finds a wireless client that's associated and communicating with an AP — and gathers the client's RF channel and MAC address information.

2. The attacker sends a Deauthenticate or Disassociate frame to the client system, forcing it to disconnect from the AP.

3. The attacker then enables a fake AP — posing as the original AP, using the same SSID and MAC address, with the only difference being that his system has to run on a different wireless channel — let's say channel 1 instead of channel 6.

4. The client system automatically tries to reauthenticate and associate itself with the original AP — only this time the odds are good that it will connect to the attacker's rogue system instead.

5. The attacker's system then connects to the original AP so all client traffic is forwarded to the victim's system — and the victim's traffic is forwarded to the rogue system.

The attacker has successfully inserted his system into the middle of the client-to-AP communications stream — and achieved "man-in-the-middle" status.

The monkey_jack utility can perform this type of wireless MITM attack. If you have the AirJack suite downloaded and compiled on a Linux-based system, the following parameters can be used to run the program:

```
# ./monkey_jack -h
Monkey Jack: Wireless 802.11(b) MITM proof of concept.
Usage: ./monkey_jack -b <bssid> -v <victim mac> -C <channel number> [ -c
            <channel number> ] [ -i <interface name> ] [ -I <interface name> ]
            [ -e <essid> ]
-a: number of disassociation frames to send (defaults to 7)
-t: number of deauthentication frames to send (defaults
to 0)
-b: bssid, the mac address of the access point (e.g.
00:de:ad:be:ef:00)
-v: victim mac address.
-c: channel number (1-14) that the access point is on,
defaults to current.
-C: channel number (1-14) that we're going to move them to.
-i: the name of the AirJack interface to use (defaults to
aj0).
-I: the name of the interface to use (defaults to eth1).
-e: the essid of the AP.
```

Now you know the parameters it requires, here's an example. We'll use `monkey_jack` to insert our system (using ports `aj0` and `eth0`) between the wireless client `00:09:5B:FF:FF:FF` and the AP `00:40:96:FF:FF:FF` with an SSID of `doh!`. We'll also force it from wireless channel 6 to channel 1, and use the defaults for all other parameters. Here we go:

```
# ./monkey_jack -b 00:40:96:FF:FF:FF -v 00:09:5b:FF:FF:FF -C 6 -c 1 -I eth0 -e
       "doh!"
```

So there you have it — assuming you received no errors during the execution of the command shown here, you're now officially the man-in-the middle.

ARP-poisoning attacks

Attackers can exploit ARP (Address Resolution Protocol) if it's running on your network. The aim is to make their systems appear to be authorized hosts on your network. What happens with this attack is that a client running a program such as `dsniff` or Ettercap can change the ARP tables — the tables that store IP addresses to MAC-address mappings — on network hosts. This causes the victim computers to think they need to send traffic to the attacker's computer (rather than to the true destination computer) when communicating on the network.

This security vulnerability is inherent in how ARP communications are handled. Compounding the problem is the fact that wireless networks use a shared medium that makes this type of attack even easier.

Walking through a typical ARP attack

Here's a typical ARP spoofing attack with a hacker's computer (`Hacky`), a legitimate wireless user's computer (`Waveboy`), and the AP (`Commander`):

1. Hacky poisons the ARP cache of victims Waveboy and Commander by using dsniff, ettercap, or a similar utility.

2. Waveboy associates Hacky's MAC address with Commander's IP address.

3. Commander associates Hacky's MAC address with Waveboy's IP address.

4. Waveboy's traffic and Commander's traffic are sent to Hacky's IP address first.

5. Hacky loads a network analyzer and captures all traffic between Waveboy and Commander. If Hacky is configured to act like a router and forward packets, it forwards the traffic to its original destination, and the original sender and receiver never know the difference!

MITM attacks that exploit ARP spoofing vulnerabilities are slightly more difficult but are still a threat. This type of attack takes advantage of the fact that ARP packets — just like 802.11 management frames — do not require any type of authentication and are easily spoofed.

An attacker can also execute a nifty *traffic-redirection attack* by using his own system as the end point. This ends up redirecting all traffic originally destined for the victim's system to the attacker's system instead. This process is depicted in Figure 12-11.

Using Ettercap

The Ettercap program can perform this type of wireless MITM attack. The following screen captures of Ettercap NG for Windows show the options for executing MITM attacks from a nice GUI interface.

1. **Load Ettercap NG and choose *Unified sniffing* from the Sniff menu.**

2. **Select the NIC you want to use from the drop-down list, as shown in Figure 12-12.**

3. **After the program loads, choose the type of attack you want to execute from the MITM menu, as shown in Figure 12-13.**

   In our example here, you'd select Arp poisoning.

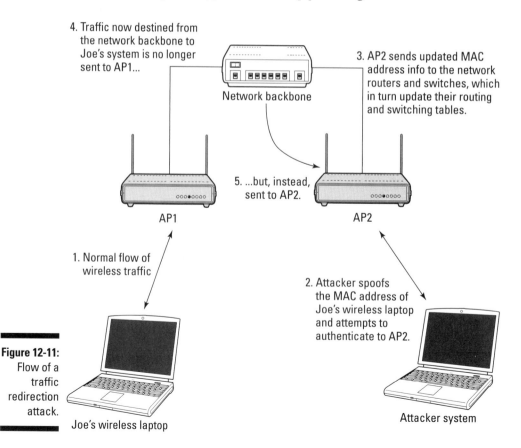

4. Traffic now destined from the network backbone to Joe's system is no longer sent to AP1...

3. AP2 sends updated MAC address info to the network routers and switches, which in turn update their routing and switching tables.

Network backbone

5. ...but, instead, sent to AP2.

AP1

AP2

1. Normal flow of wireless traffic

2. Attacker spoofs the MAC address of Joe's wireless laptop and attempts to authenticate to AP2.

Figure 12-11: Flow of a traffic redirection attack.

Joe's wireless laptop

Attacker system

Figure 12-12:
Selecting
a NIC for
Ettercap
NG to use.

Figure 12-12:
Selecting
a NIC for
Ettercap
NG to use.

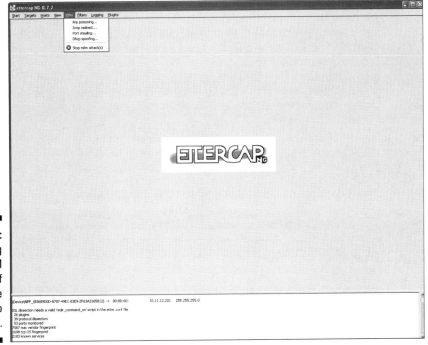

Figure 12-13:
Selecting
the MITM
attack of
your choice
in Ettercap
NG.

Again, note how simple it is to achieve a MITM attack. At this point, you can use Ettercap NG and your favorite network analyzer to capture your victim system's data — or launch other attacks of the type mentioned in this section.

SNMP: That's Why They Call It Simple

Simple Network Management Protocol (SNMP) is a protocol built in to virtually every network infrastructure device — both wireless and wired. Everything from switches to routers to servers to APs can be managed via SNMP. There are various network-management programs such as HP OpenView (www.managementsoftware.hp.com), LANDesk (www.landesk.com), and Silverback Technologies (www.silverbacktech.com) that use SNMP for remote network-host management. Their capabilities are especially helpful in wireless networks when you're trying to manage what's happening on your airwaves. Unfortunately, they all depend on SNMP — which presents various security vulnerabilities.

The problem is that most wireless APs run SNMP as is — not locked down from the elements. In fact, most APs have SNMP enabled when it doesn't need to be. If SNMP is compromised, a hacker can gather network information and use it to attack your systems. If a hacker is trying to attack your wireless network and SNMP shows up in her port scans, you can bet she'll try to compromise the system.

Figure 12-14 shows how GFI LANguard Network Security Scanner was able not only to detect that SNMP is enabled on a Cisco Aironet AP but also to glean some basic information from it.

In Figure 12-15, the QualysGuard vulnerability-assessment tool discovered that this same AP has writeable SNMP information due to an insecure SNMP community name. This could be especially bad if you're trying to manage such an AP and an attacker is able to modify its settings!

If you want to perform a quick-and-dirty test to see whether SNMP is running on a host, perform a port scan and look to see if UDP port 161 is open. If it is, then SNMP is alive and well — and vulnerable — on the host system.

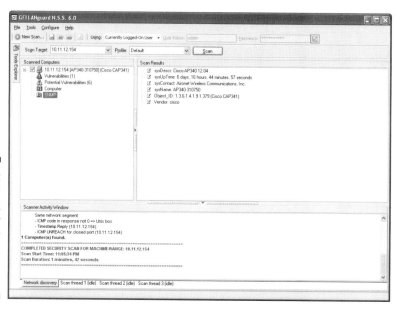

Figure 12-14:
LANguard
Network
Security
Scanner
discovered
SNMP
information.

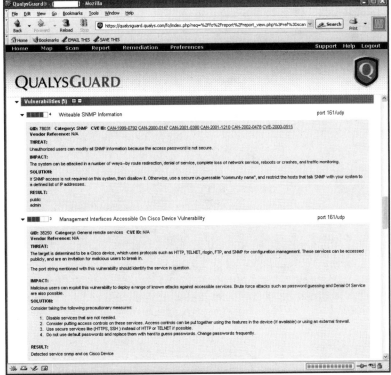

Figure 12-15:
Qualys-
Guard
discovered
that SNMP
information
is writable.

Various other utilities — both Windows- and UNIX/Linux-based — can enumerate SNMP on APs and other wireless hosts:

✔ Windows GUI-based Getif (`www.wtcs.org/snmp4tpc/getif.htm`)

✔ Windows text-based SNMPUTIL (`www.wtcs.org/snmp4tpc/FILES/Tools/SNMPUTIL/SNMPUTIL.zip`)

✔ UCD-SNMP (`www.ece.ucdavis.edu/ucd-snmp`)

If you have APs with default SNMP enabled on your wireless network, the best-case scenario is that an attacker will be able to enumerate those systems and glean AP information such as system uptime, hardware model number, and firmware revision as shown in Figure 12-16. And that's a *best* case.

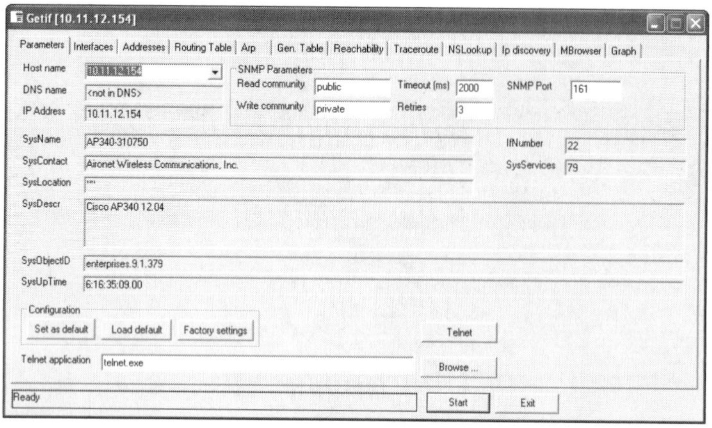

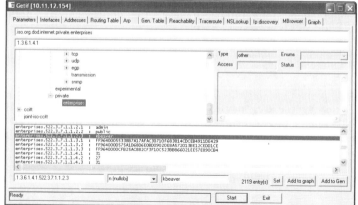

Figure 12-16:
General
SNMP
information
gather using
getif.

An attacker can use `getif` or similar tool to glean information such as MAC addresses that have associated with the AP — and even snag AP usernames for HTTP management, as shown in Figure 12-17.

This information is certainly not what you need to be advertising to the outside world. But you knew that. What you may not have known is that it's already out there.

The worst-case scenario is that you'll have one or more APs running a seriously vulnerable implementation of SNMP version 1 that can lead to DoS attacks, unauthorized access, and more. For a list of vendors and products that are affected by the well-known SNMP vulnerabilities, check out `www.cert.org/advisories/CA-2002-03.html`.

Figure 12-17:
HTTP user
IDs gleaned
via getif's
SNMP
browsing
function.

All Hail the Queensland Attack

A relatively new attack against the 802.11 protocol showed up Down Under in May 2004, discovered by researchers at Queensland University of Technology's Information Security Research Centre (www.kb.cert.org/vuls/id/106678) in Australia. This attack, initially referred to as the Clear Channel Assessment attack, affects the Direct Sequence Spread Spectrum function that works as part of 802.11's Carrier-Sense Multiple Access/Collision Avoidance (CSMA/CA) protocol that manages the wireless communications medium. This attack is often called the *Queensland Attack* — crediting the researchers who discovered it.

Wireless systems (clients, APs, and so on) use CSMA/CA to determine whether or not the wireless medium is ready and the system can transmit data. The Queensland attack exploits the Clear Channel Assessment (CCA) function within CSMA/CA and basically makes it appear that the airwaves are busy — effectively preventing any other wireless system from transmitting. This denial of service is accomplished by placing a wireless NIC in continuous transmit mode.

With the right tool, the Queensland Attack is relatively simple to execute. It can wreak havoc on a wireless network, effectively bringing it to its knees. There's very little that can be done about it, especially if the attacker's signal is more powerful than that of your wireless systems. That's no problem for hackers equipped with a high-powered wireless NIC combined with a high-gain antenna (see Chapter 13 for more information). Combine an easily over-powered network with the fact that 802.11 systems use a shared medium to communicate, and you have the makings of a very effective attack.

All it takes for an attacker to run such an attack against your wireless systems is to run an old Prism chipset-testing program called Prism Test Utility (PrismTestUtil322.exe). This program was previously available for public download on Intersil's Web site — and it's still easy to find elsewhere with a basic Internet search, so it's probably not going away any time soon. This attack can just as easily be carried with other hardware tweaking as well.

Although the Queensland Attack exploits an 802.11 protocol issue, it could just as easily be considered a DoS attack, given its outcome (big-time denial of service). Refer to Chapter 13 for an in-depth look at various wireless DoS attacks.

Sniffing for Network Problems

As we've demonstrated in various other chapters in this book, a wireless network analyzer (sniffer) is a tool that allows you to look into the network and analyze data going across the airwaves for network optimization, security, and/or troubleshooting purposes. Like a microscope for a lab scientist, a wireless network analyzer is a must-have tool for any security professional performing ethical hacks against wireless networks.

A network analyzer is just software running on a computer with a network card. It works by placing the network card in *promiscuous mode,* which enables the card to see all the traffic on the network, even traffic not destined to the network analyzer host. The network analyzer performs the following functions:

✔ Captures all network traffic

✔ Interprets or decodes what is found into a human-readable format

✔ Displays it all in chronological order

There are literally dozens of neat uses of a wireless sniffer beyond capturing cleartext communications and searching for SSIDs. Such a program can help with:

✔ Viewing anomalous network traffic and even tracking down intruders.

✔ Developing a baseline of network activity and performance before a security incident occurs.

The next section outlines specific network information to look for.

Network-analysis programs

You can use one of the following programs for network analysis:

✔ **AiroPeek and AiroPeek NX by WildPackets (`www.wildpackets.com`):** It delivers a ton of features that the higher-end network analyzers of yesterday have — for a fraction of their cost. AiroPeek is available for the Windows operating system.

✔ **CommView for WiFi (`www.tamos.com/products/commwifi`):** Again, very feature-rich, especially given its low price. It also includes a packet generator that can really come in handy. See Chapter 13 for more details on using this feature of CommView for WiFi. CommView for WiFi is available for the Windows operating system.

✔ **AirMagnet Laptop Analyzer** (`www.airmagnet.com/products/laptop.htm`): This program is great for wireless security testing as well. It has a great user interface and is very easy to use. AirMagnet Laptop Analyzer is available for the Windows operating system.

✔ **AirDefense Mobile** (`www.airdefense.net/products/admobile`): Similar to each of the programs in this list, AirDefense Mobile offers a wide range of features, all within an easy-to-use GUI interface. AirDefense Mobile is available for the Windows operating system.

✔ **Ethereal** (`www.ethereal.org`): Ethereal is a great open-source (free) program, especially if you need a quick fix and don't have your test system nearby. It's not as user-friendly as many other programs, but it is very powerful if you're willing to learn its ins and outs. Ethereal is available for both Windows- and UNIX-based operating systems.

A slew of other wireless network analyzers are available as well, including Kismet, many of which we cover in other chapters. A general rule of thumb is that you get what you pay for. Don't worry about whether you're using the *right* network analyzer. The *right* network analyzer is the one that works best for you — the one that feels the most comfortable and the one that does what you need it to do — after you've done some careful experimenting.

Network analyzer tips

Before getting started, configure your network analyzer to capture and store the most relevant data. If your network analyzer permits it, configure your network analyzer software to use a *first-in, first-out buffer.* This overwrites the oldest data when the buffer fills up, but it may be your only option if memory and hard-drive space are limited on your network-analysis computer.

Also, if your network analyzer permits it, record all the traffic into a capture file and save it to the hard drive. This is the ideal scenario — especially if you have a large hard drive (50GB or bigger).

You can easily fill a several-gigabyte hard drive in next to no time, so don't capture all packets unless absolutely necessary.

Often the most practical way to use a network analyzer is to just let it run in *monitor mode* if your analyzer supports it — capturing overall statistics of the network (SSIDs, channels used, active nodes, protocols seen, and so on) without capturing every single packet. You can often glean enough information from a network analyzer's monitor mode to look for security weaknesses. Just keep in mind that you may need to let your network analyzer run for quite a while — from a few minutes to a few days — depending on what you're looking for.

Weird stuff to look for

A network analyzer is one of the best security tools you can own. It's amazing what you can find on your network that you wouldn't know about otherwise (and *really need to* know about). The following list sums up various types of traffic and trends you can look for to help you find security vulnerabilities in your wireless network.

✔ Protocols in use:

- Non-standard or unsupported traffic such as instant messaging, POP3 e-mail, FTP, and telnet.

- ICMP packets — especially in large numbers — which could indicate potential ping sweeps for the start of system enumeration.

✔ Usage trends:

- What are your peak wireless network usage times?

- Are you seeing heavy traffic during off peak hours?

- Internet usage habits can help point out malicious behavior of a rogue insider or a system that has been compromised.

AirMagnet Laptop Analyzer's Channel monitor (as shown in Figure 12-18) is great for observing wireless trends over time.

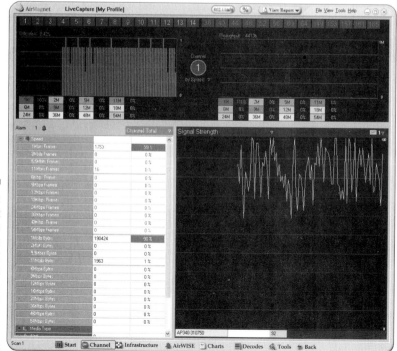

Figure 12-18:
AirMagnet
Laptop
Analyzer's
Channel
view
showing
graphical
usage
trends.

✔ MAC addresses:

- Do you know which MAC addresses belong on your network?

- Look for odd, default, and duplicate MAC addresses.

If you spot an odd MAC address and have CommView for WiFi, you can perform a quick lookup using the built-in NIC Vendor Identifier available from the Tools menu as shown in Figure 12-19.

Figure 12-19: CommView for WiFi's NIC Vendor Identifier.

This comes in handy if you know you only use a certain vendor's NICs — and spot an odd one on your network.

CommView for WiFi's NIC Vendor Identifier utility is especially useful if you don't have access to the Internet to perform a lookup because your wireless-security programs have control of your wireless NIC.

✔ Network errors and anomalies:

- CRC errors

- WEP errors

- Excessive amounts of oversized packets

- Excessive amounts of multicast or broadcast traffic

- Excessive DHCP requests

- Excessive retries

Discovering these types of network issues is made simple by AiroPeek's Summary page, as shown in Figure 12-20.

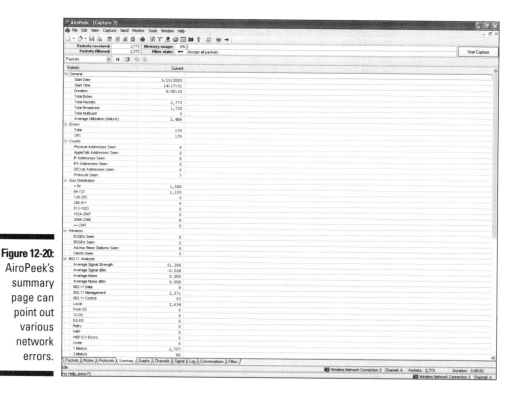

Figure 12-20:
Figure 12-20:
AiroPeek's
summary
page can
point out
various
network
errors.

Network Attack Countermeasures

There are various countermeasures you can put in place to defend against many of the network-level attacks we've outlined in this chapter.

Like with all other wireless-network countermeasures, never assume that the lower layers of your wireless network (Physical Layer 1 and MAC Layer 2) are secure just because you have high-layer security mechanisms in place (such as firewalls and authentication systems).

The following are effective countermeasures against wireless-network attacks:

- ✔ Enable WEP, WPA, or use a VPN to protect wireless communications.
- ✔ Disable SNMP if you are using it to manage your network.
- ✔ Change your SNMP community string if you do use it.

✔ Disable other protocols and services you don't need on your wireless infrastructure systems (such as ICMP, telnet, and HTTP).

✔ Segment your wireless systems away from your wired network — preferably in a DMZ off your perimeter firewall.

✔ Utilize switch-based port security to ward off ARP attacks.

✔ Use the directional antennae and AP power settings where possible to help keep your signals out of unfriendly airwaves.

✔ Use a wireless IDS/IPS system to monitor your airwaves and ward off network attacks.

Chapter 13

Denial-of-Service Attacks

. . . .

. . . .

*I*magine experiencing all of the following scenarios simultaneously:

  ✔ You're *trying* to finish a presentation you have to give in 30 minutes.

  ✔ You're on a conference call you were forced to participate in at the last minute.

  ✔ Your other office phone line is ringing.

  ✔ Your cell phone is jamming out "Another One Bites the Dust," signifying yet another call from an agitated user.

  ✔ Your organizer software is alerting you of a meeting you're about to miss.

  ✔ Someone's waiting outside your cubicle with a delivery.

  ✔ Your PDA has vibrated itself out of its cradle and onto the floor from all the e-mail alerts you're getting from your firewall.

  ✔ Your computer crashes with a Blue Screen of Death, just before you have any chance to save your Power Point file.

  ✔ Your colleague in the next cubicle is asking you a question.

  ✔ The building fire alarm starts going off.

This may be the closest you'll get to experiencing a personal *denial-of-service (DoS)* attack. In other words, an "attack" in which an overwhelming number of circumstances, most of which are beyond your control, prevent you from focusing on the task at hand and getting your work done.

DoS attacks against your wireless network aren't much different. In essence, they send a horde of malicious network requests — or overload the airwaves and wireless systems with junk traffic — preventing legitimate requests from being addressed. This Achilles heel of wireless networks can affect your systems in ways you may have never imagined, leaving your systems completely defenseless. No mystery that DoS attacks just happened to fall into the "Number 13" bad-luck chapter.

Many of the recent, highly publicized hacker attacks against popular Web sites and e-commerce companies have been DoS attacks. These were carefully crafted attacks — often utilizing *thousands* of compromised systems — that were able to bring down Web servers from across the Internet. It's worth noting, however, that such attacks are not all that common against wired networks; that's because they typically require a high-level of skill and planning to carry out. With DoS attacks against wireless networks, however, we're not so lucky — ordinary levels of hacker competence can produce way too much network mayhem.

Most network DoS attacks are performed out of pure malice — and often for the fun of it — but sometimes they're performed for competitive or political purposes. Often these attacks aim to further penetrate a network or force an administrator to try different security mechanisms (or none at all) — even while troubleshooting and trying to find out why signals and systems are dropping like flies. The typical motivation, however, is more basic — to take network service away from others and keep them from doing what they need to do.

Here's where the notorious vulnerabilities of the IEEE 802.11 specifications (gruesomely detailed in Chapter 12) come into play. 802.11-based systems — including wireless clients, access points, and the entire radio spectrum they operate over — can be completely compromised in a much simpler fashion. All it takes is a few basic tools and minimal know-how to perform some wicked DoS attacks against wireless systems — not necessarily what you bargained for when you implemented your way-cool and convenient wireless network.

There are two main reasons that 802.11-based wireless systems are vulnerable to DoS attacks:

> ✔ **Lack of frame authentication in 802.11 management frames such as beacons, association requests, and probe responses.** The functionality inherent in the MAC layer of a 802.11-based network is all about access: It allows wireless systems to discover, join, and basically roam free on the network, completely exposed to the elements. This implicit trust among wireless systems makes it easy for attackers to spoof legitimate devices and bring down individual hosts — or even an entire wireless network — all at once.

✔ **Lack of physical boundaries for radio waves.** Radio is everywhere, and can come from anywhere. This makes attacks simpler and reduces the likelihood that an attacker will get caught. Additionally, APs and other wireless infrastructure equipment are often exposed in easy-to-access areas where they're more susceptible to tampering and theft.

You can easily create self-inflicted DoS conditions on your systems when you test for such vulnerabilities. Running the wrong tools — or the right tools without understanding (and being prepared for) their consequences — can crash your wireless network or cause your data to be corrupted or compromised. Such "results" are certainly one quick way to get on the bad side of a lot of people. Be careful when you use any of the tools we mention or demonstrate on this chapter, starting with this rule. *Always test your tools on non-production systems first if you're not sure how to use them.* Such precautions help prevent DoS conditions that could disrupt your live systems.

Given the danger involved in performing DoS tests against your own systems, this chapter will be a little different from other ones in this book: It's more about attack *education* than attack demonstration. We outline the various types of DoS attacks against wireless networks, and then show you what some DoS tools can do if you choose to perform such attacks against your systems.

What Can Happen

Denial of Service attacks do just that — deny service. They prevent legitimate wireless users and systems from performing typical tasks such as

✔ Connecting to the wireless network

✔ Staying connected to the wireless network

✔ Serving up various network requests

✔ Managing network communications

Obviously, disruption of these types of network services can wreak havoc on usability — and can even threaten data integrity.

Types of DoS attacks

DoS attacks can come at various levels within a wireless network. They can impact radio signals, network protocols, and even wireless applications. Signals can be jammed, wireless devices can be spoofed so the bad guys can perform malicious acts, and APs can be overloaded. In addition, if vulnerable APs and ad-hoc clients are located behind the network firewall and are attacked or somehow compromised, there's a chance that the *wired* network can be negatively affected.

Wireless attackers can even take advantage of vulnerabilities in the power-saving features of client computers. Here are some typical gambits:

- ✔ Tricking an AP into thinking that a specified client is going to sleep — when it's not — which stops the client from transmitting and receiving packets.

- ✔ Spoofing a wireless client to make an AP think that a client has awakened from its power-saving sleep — when it hasn't. The AP thinks the client is ready to receive packets that have queued up to wait for its attention, and sends the packets. Result: traffic jam.

- ✔ Forcing a wireless client to stay asleep — which keeps it from communicating on the network.

- ✔ Preventing a wireless client from going to sleep — potentially causing its battery to run down prematurely.

Two highly popular attacks that could be categorized as denying service — hijacking and MITM attacks — come at the network level. (These are covered in Chapter 12.) DoS attacks can even be accomplished by exploiting the weaknesses in encryption and authentication algorithms such as WEP and WPA. We cover these attacks in Chapters 14 and 15, respectively.

It's so easy

DoS attacks, especially those that come at the RF level — riding the radio beam in — are ridiculously easy to carry out for several reasons:

- ✔ Physically separating potential attackers from the radio waves they're trying to use can be very difficult and costly.

- ✔ Unlike military and other custom wireless applications, commercial applications are quite commonplace. The bad guys have the same commercial equipment we do.

- ✔ An attacker can increase the DoS capabilities of a rogue system simply by increasing the RF transmitting power.

- ✔ The 802.11 wireless protocols were designed for usability and compatibility — not necessarily to protect against DoS attacks.

Unfortunately, the defenses for DoS attacks are few and far between. It really pays to be proactive and understand what can happen and, if you so desire, perform your own DoS testing to see how vulnerable your systems are *before* a problem occurs.

DoS attacks against wireless systems are not only difficult to prevent but hard to trace; it can be next to impossible to determine where the attacks are coming from. This is why you've got to slip into the mindset of the attacker and test your own systems, or at least implement reasonable countermeasures to keep the predators at bay.

Before you start regretting ever venturing into the wireless arena, keep in mind that many DoS attacks are purely theoretical (less practical) in nature and have no supporting tools and no confirmed existence in the real world. We don't mean that DoS attacks on wireless networks shouldn't be taken seriously; we're just saying the picture is not as bleak as many make it out to be. Accordingly, we'll stick to the more practical DoS attacks and tests in this chapter. Let's jump right into things.

We Be Jamming

As you might expect, a major type of DoS attack that wireless networks are susceptible to is RF jamming. Wireless network signals can be disrupted and prevented from doing their work (jammed) when another radio signal that operates in the same or nearby frequency range. The normal ranges for 802.1*x*-based network communications are 2.4 GHz (for 802.11b and g) or 5 GHz (for 802.11a). A high-powered rogue signal can interferes with — or overpower — the network's existing radio transmissions. Technically, the Queensland Attack covered in Chapter 12 could be considered a type of jamming attack as well as a DoS attack.

Wireless networks are very sensitive to jamming because of their low-power operation and the relatively narrow bandwidth (22 MHz per channel) they use to communicate. Depending on the power of an incoming rogue signal, jamming and other RF interference can cause your systems to drop a few packets here and there — or create complete communications breakdown. Both effects can be equally disruptive.

Unlike 802.11b and g networks, which use the crowded ISM band, 802.11a equipment is much less susceptible to jamming caused by interference from other devices because it runs in the 5 GHz frequency range.

RF jamming can occur unintentionally from nearby equipment. It can also occur maliciously by an attacker with an RF jammer, a wireless laptop with a high-powered NIC such as the 300mW PC Card NIC sold by Demarch Technology Group (www.demarctech.com), or even a high-powered AP. A high-gain or directional antenna that can boost the attacker's signals can wreak greater havoc. This will not only increase the power output but will also provide the added benefit of physical distance between the attacker and the system she's jamming.

RF jamming can force wireless clients to roam the available frequencies, searching for an alternate access point to communicate with — and they may find one on somebody else's available wireless network. When a client finds an alternate AP, it may inadvertently authenticate and associate to one of your own APs — or (worse) to a rogue AP that the attacker has set up. We talk more about this type of commandeering in Chapter 12.

Common signal interrupters

Various types of radio-transmission devices can disrupt 802.11-based wireless networks — especially 802.11 b and g systems that operate in the ISM band. The interesting thing is that many such troublesome devices are common everyday electronics present in our offices and homes — these, for instance:

- 2.4 GHz cordless phones
- Wireless security cameras
- Bluetooth systems
- Baby monitors
- Microwave ovens
- Radio power generators (more on these below)
- X-10 home automation equipment

The reason that these devices are all here in the 2.4 GHz spectrum is that these are all low-powered RF devices that can be operated without the owner or operator requiring a license from the Federal Communications Commission (FCC). All of these devices are capable of causing wireless-network disruption that can lead to intermittent network connectivity or (worse) self-inflicted DoS attacks that you didn't intend to create.

What jamming looks like

Before we get too far into jamming attacks, it makes sense to show you what RF interference actually looks like when it's happening.

Figure 13-1 shows what a strong 802.11b signal looks like in NetStumbler's Channel view. Note that although this is not a true RF spectrum analyzer (which can show detailed radio-frequency information), we can still see the signal disruption taking place. The tall and even bars shown in the figure represent a strong and continuous signal.

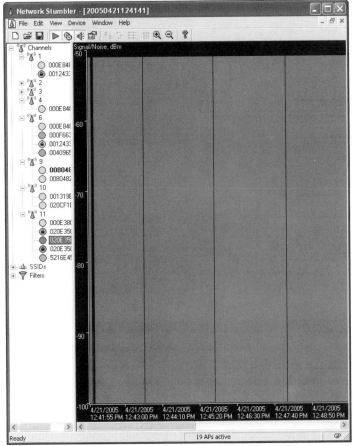

Figure 13-1:
Strong
wireless
signal
experi-
encing no
interfer-
ence.

Figure 13-2 shows an 802.11b signal that's experiencing some random noise and signal loss. Notice the signal profile: It's degraded and choppy compared to that of Figure 13-1.

Figure 13-3 shows an 802.11b signal that's experiencing severe jamming. Notice that although the signal is strong at times, it's missing across various time periods and is being overpowered by another signal. This secondary signal is shown in red at the bottom of the green (actual) signal in NetStumbler. NetStumbler also shows a purple bar that signifies a potential loss of radio signal.

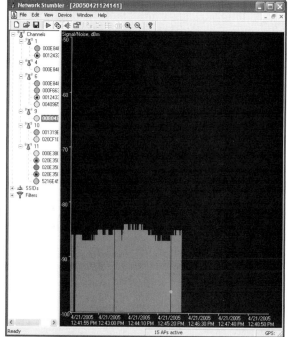

Figure 13-2:
Wireless
signal
experi-
encing mild
interfer-
ence.

Fight the power generators

As we alluded to earlier, DoS attacks against 802.11 wireless systems can also be carried out through the use of RF jammers — also known as *radio power generators* or *signal generators.* Most companies don't sell devices called RF "jammers"; instead, they market them as signal generators for the purpose of designing and testing radio signals, cabling, antennae, and so on.

Such devices can generate power levels that range from several hundred milliwatts up to several watts, across broad frequency ranges — easily overpowering the weaker 802.11 signals that usually run in the low end of the 1-to-100 milliwatt range. If you're into electronics design (and have the know-how and parts), you can make your own radio-power-generator system. Thankfully, for those of us who don't have that kind of time or patience, several commercial signal generators are available. They're helpful tools for testing your wireless network's susceptibility to DoS attacks when it's subjected to such powerful signals. Two systems we're familiar with are the following:

✔ YDI Wireless (now Terabeam Wireless) makes the PSG-1 signal generator (www.ydi.com/products/test_eq/psg.php)

✔ Global Gadget offers the 2.4JM signal generator (www.globalgadgetuk.com/wireless.htm)

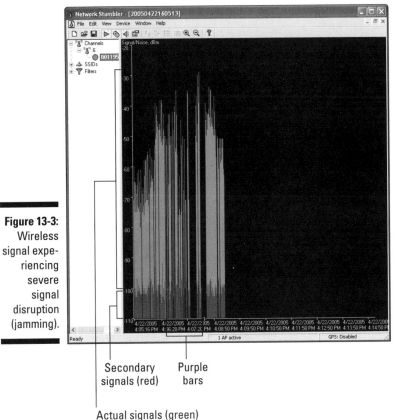

Figure 13-3:
Wireless signal experiencing severe signal disruption (jamming).

Secondary signals (red)   Purple bars

Actual signals (green)

TIP

There are also various signal-generator vendors listed at online e-commerce sites such as Naptech (www.naptech.com) and TestMart (http://signal generator.testmart.com).

A jamming attack against a wireless network can be carried out from several dozen meters away, which helps the attacker hide. The two jammers we mentioned are handheld systems — so an attacker could conceivably have one stored in his pocket or briefcase, and you'd be none the wiser. Perhaps the most frustrating thing about jammers is that even the most highly protected wireless systems are pretty much indefensible in the face of such an attack.

We won't demonstrate what using a radio power generator can do to a wireless network — but suffice it to say that the outcome is likely to be worse than the RF signal disruption shown earlier in Figure 13-3.

AP Overloading

802.11-based wireless access points can only handle so much traffic before their memory fills up and their processors become overloaded. This type of DoS attack overloads not only the wireless medium (as outlined earlier) but also the actual wireless infrastructure — and APs themselves.

There are several ways that APs can become overloaded and simply stop addressing the needs of existing or new clients — or just break down altogether. Some of these de-facto attacks are unintentional; others are deliberate and malicious. Let's take a look at what can happen.

Guilty by association

Attackers can exploit a weakness in the way access points queue incoming client requests — beginning with the *client association identifier (AID) tables* — the section of an AP's memory that stores client connection information. The AID tables only have a finite amount of memory and thus can only handle a limited number of wireless client connections. Once this memory fills up, most APs will no longer accept incoming association requests; some APs even crash.

These types of DoS attacks typically use one of two methods:

- ✔ Association flooding
- ✔ Authentication flooding

Both are easier to do when anybody can connect. When APs are set up to use "open" as the default authentication type, just about any client (trusted or untrusted) can connect to the AP. This is one of those fundamental 802.11 security flaws deemed necessary to keep wireless-connectivity headaches to a minimum. Such *open authentication* allows any client to send two critical requests:

- ✔ Authentication requests for initial connectivity
- ✔ Association requests to "join" the wireless network

Now, wireless client connectivity to an AP that's running open authentication has the three basic phases:

1. No connection

2. Authenticated but not associated

3. Authenticated and associated

This three-step process is critical for understanding DoS attacks, so we show it again in Figure 13-4.

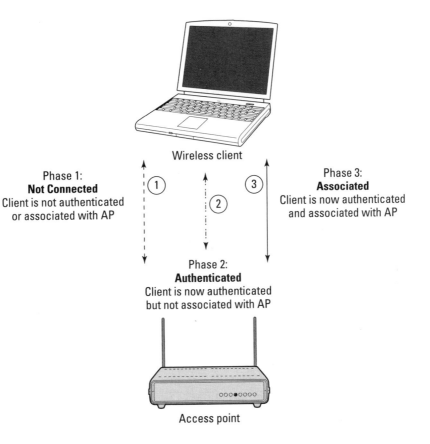

Figure 13-4:
Client-to-AP
connection
process.

Wireless client

Phase 1:
Not Connected
Client is not authenticated
or associated with AP

Phase 3:
Associated
Client is now authenticated
and associated with AP

Phase 2:
Authenticated
Client is now authenticated
but not associated with AP

Access point

Attacks that overload the AID tables create a situation that can take a wireless network from normal to frozen in no time: Even an average number of legitimate wireless-client connections can multiply to an insane number when illegitimate connections pile on, faster than you can say *intrusion prevention.*

Association and authentication attacks are possible mainly because 802.11 management-frame requests and sequencing are not authenticated — or monitored for anomalies.

If you're up for testing to see how easy it is to fill up the AID tables on your AP(s), there are several tools you can use. One of our favorites is Void11 — a packet-injection tool. Figure 13-5 shows its options: Notice the authentication- and association-flood options, as well those for flooding a single target, broadcast systems, and randomly generated systems.

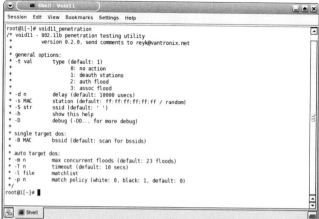

Figure 13-5:
The various
options of
the Void11
packet-
injection
tool.

You can download Void11 from the WLSec project homepage at www.wlsec.
net/void11. Or, if you're not too fond of trying to get your wireless NIC to
work in UNIX/Linux, you can run Void11 directly off the super-cool KNOPPIX
CD-ROM-based Auditor Security Collection (http://new.remote-exploit.
org/index.php/Auditor_main). See Chapter 15 for more details on using
and tweaking the Auditor hacking tools.

A great Windows-based tool for creating association and authentication
attacks is CommView for WiFi's Packet Generator Tool, shown in Figure 13-6.

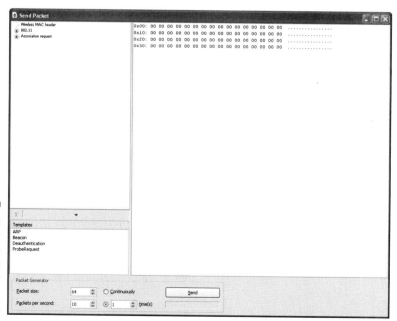

Figure 13-6:
CommView
for WiFi's
Packet
Generator
tool.

Packet Generator, which is very easy to use, allows you to replay practically any 802.11 packet (including Association and Authentication Request packets) that you've captured in CommView for WiFi or another network-analyzer program.

Here's a brisk walkthrough capturing an association request packet in CommView for WiFi, copying the packet to the Packet Generator tool, and then sending the packet onto the airwaves:

1. **Load CommView for WiFi and click the blue Start Capture icon in the upper-left corner or simply press Ctrl+S on your keyboard.**

   This loads the Scanner utility (as shown in Figure 13-7) so you can enable your wireless NIC to capture packets.

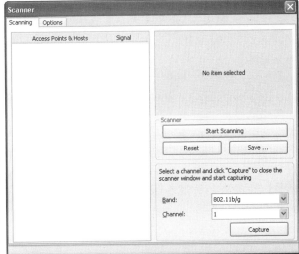

Figure 13-7:
CommView
for WiFi's
Scanner
utility.

2. **Click the Capture button on the Scanner window.**

   This "opens" the Wireless Adapter Enable Promiscuous mode on your wireless NIC, and allows you to start capturing wireless packets.

3. **Capture an Association Request packet.**

   The easiest way to do this is to power on a new wireless client and look for its requests to the AP to associate. Packet number 115 in Figure 13-8 shows what an Association Request packet looks like. Note that CommView for WiFi lists this as a MGNT/ASS REQ. packet where the MGNT represents a *management* type packet.

4. Copy the Association Request packet into Packet Generator.

You can do this by following these steps:

   a. Ensure you have the packet you wish to copy highlighted and then press Ctrl+R to load the Packet Generator tool.

   b. Within the Packet Generator window, click the black Up arrow next to the sigma (Σ) symbol to show the Templates section.

   c. Resize both the CommView for WiFi window and the Packet Generator window so you can view both on your desktop.

   d. Simply drag and drop the Association Request packet into the Templates section of the Packet Generator window.

5. Rename the packet.

In the Packet Generator tool, simply right-click the packet labeled New Template(0) in the Templates section and enter a new name such as AssociationRequest. Click outside of the name area to make the change permanent.

That's all there is to it! You're now ready to use CommView for WiFi's Packet Generator tool to send the Association Request packet to your AP(s). Note that if you'd like to change the source or destination MAC addresses in the packet, you can do so very easily by simply clicking into the hex-data area of the Packet Generator window and changing the data directly. (We walk you through this process later in the chapter, in the section called "Deauthentications.")

6. Send the packet.

You can send the packet by simply selecting the AssociationRequest (or whatever you named it) packet in the Templates section and clicking Send in the Packet Generator tool. Note that you can change the packet size, number of packets per second, and the number of times to send it.

This exercise demonstrates how simple it is to create an association-flood attack. This whole process (depicted in Figure 13-9) — and its potentially harmful results — can happen in a split second.

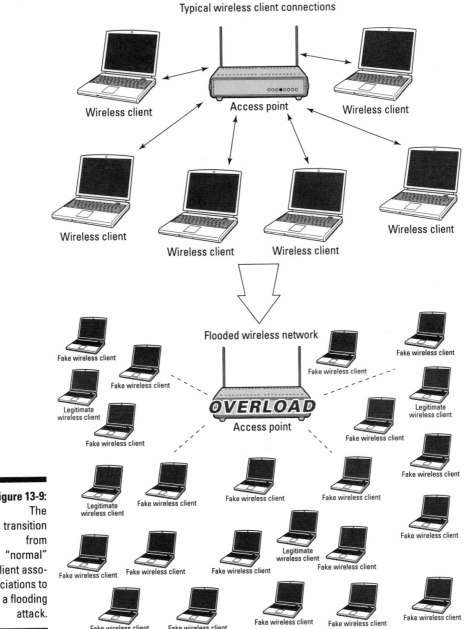

Figure 13-9:
The transition from "normal" client associations to a flooding attack.

The same test can be performed with Authentication Request packets as well.

We'll use CommView for WiFi's Packet Generator tool again when we look at deauthentication and disassociation attacks later in this chapter. We'll also demonstrate what such attacks look like through a network analyzer.

Other packet injection tools can be used to execute association-flooding attacks if you're eager to venture out, including the following UNIX/Linux-based tools:

- file2air (`http://home.jwu.edu/jwright/code/file2air-0.1.tar.bz2`)
- AirJack (`http://sourceforge.net/projects/airjack`)
- libradiate (`www.packetfactory.net/projects/libradiate`)

Too much traffic

Wireless overloading is often unintentional, especially with today's "robust" applications sucking up every available bit of memory, processor time, and network bandwidth. For example, the following *legitimate* wireless network traffic is quite possible on a typical network at any given time:

- Movie and music file downloads
- Basic Web browsing
- P2P file sharing traffic
- A bored employee hosting his own Web or FTP server
- Users streaming the audio of their favorite radio talk-show host
- Internal network file copies, print jobs, and so on
- Vulnerability-assessment software running an obscene number of tests every second
- Downloads occurring over a very-high-speed Internet connection (think T3 and faster)
- Web, e-mail, FTP, or other servers transmitting and receiving data

Wireless networks can easily be saturated at speeds much lower than their claimed *throughput rate* (in effect, how fast they can transfer data). This is especially true for 802.11b systems that not only struggle to provide enough usable throughput but are also half-duplex (one side communicates at a time). This means that even in a perfect world, 802.11b systems can't obtain more than 5.5 Mbps of throughput — usually less, given the speed loss that comes from handling protocols and the traffic generated by multiple clients on the network.

A neat commercial security-testing tool you can use to test an AP's suscepti-bility to information overload is BLADE Software's IDS Informer program (www.bladesoftware.net). This software is designed for testing IDS/IPS sys-tems but can be used to flood a wireless network for DoS testing purposes just as well.

All it takes is one computer, generating a fair amount of legitimate traffic, to bring down an AP. In fact, according to previous nonscientific studies of 802.11b capabilities that Kevin was involved with, a typical 802.11b AP can handle only a dozen or so (often fewer) client connections before perfor-mance starts degrading for everyone on the network. This can occur even if the network uses multiple APs in ESS mode to service a broad wireless cover-age area. Using 802.11g systems won't necessarily fix this issue; the trouble may be simply less noticeable, camouflaged by the 54 Mbps throughput of 802.11g systems (compared to only 11 Mbps in 802.11b systems).

All of this is with *legitimate* traffic on the network. Imagine what can happen when multiple computers are generating *malicious* traffic! At best, it's cer-tainly enough to create a serious DoS condition. Technically, such an attack could be considered a *distributed DoS (DDoS) attack* because multiple sys-tems are involved.

Like their 802.11b predecessors, newer 802.11g systems can handle only three non-overlapping channels (1, 6, and 11); available bandwidth is still minimal on congested networks. This problem can be overcome by using 802.11a tech-nology, which has more available channels for communication — and allows the grouping of more APs to handle the extra requests. But do you really want to purchase and implement the Betamax of wireless network technologies?

Are You Dis'ing Me?

Several clever DoS attacks against wireless clients are bad enough to make you want to stick with good old-fashioned Ethernet — maybe even Token Ring. These attacks are often more effective than association and authentica-tion attacks — that's because wireless clients tend to be more willing to believe that anything coming to them from an AP *must* be valid.

There are two main types of DoS attacks against client systems:

- Disassociation attacks
- Deauthentication attacks

The bad thing about these types of client DoS attacks is that they can go on indefinitely until the attacker stops the attack.

Several hacking tools are available to execute client DoS attacks, including WLAN-jack (if you're lucky enough to have downloaded it before it was taken offline), Void11 (`www.wlsec.net/void11`), and FATA-jack (`www.security wireless.info/public/wipentest/fata_jack.c`). The same results can be accomplished very easily with CommView for WiFi's Packet Generator as we'll demonstrate shortly.

Disassociations

A *disassociation attack* is essentially a wireless station's way of saying "I don't want to talk to you any more." The situation is similar to when a friend ticks you off — you (the AP) tell the friend (the wireless client) to get lost. Disassociation packets can be sent from a wireless client to an AP as well.

The way a disassociation attack works is actually very straightforward. This attack simply mimics valid disassociation frames originating from a client or AP and cuts off the association. First, the attacker spoofs either the client or the APs MAC address (usually the latter). Then he sends forged disassociation packets to either a specific system or to the broadcast address. A disassociation attack is shown graphically in Figure 13-10.

After the disassociation occurs, the client is returned to a state where it's still authenticated to the AP, but not associated. This leaves it in a disconnected state from the network.

Deauthentications

A *deauthentication attack* is actually a little more effective than a disassociation attack because it puts the client in a state of complete disconnection. The deauthentication attack is a wireless station's way of saying "Your connection to me is no longer valid." As with disassociation attacks, this attack can originate at the client; otherwise the AP can be directed to an individual MAC address or the broadcast address.

Figure 13-11 shows how a deauthentication attack is carried out.

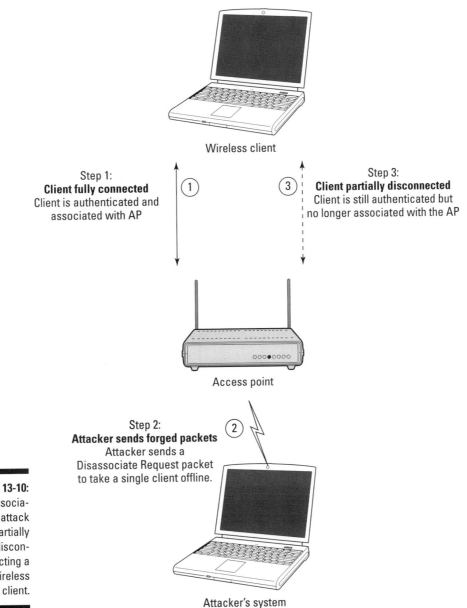

Wireless client

Step 1:
Client fully connected
Client is authenticated and
associated with AP

Step 3:
Client partially disconnected
Client is still authenticated but
no longer associated with the AP

Access point

Step 2:
Attacker sends forged packets
Attacker sends a
Disassociate Request packet
to take a single client offline.

Figure 13-10:
Disassocia-
tion attack
partially
discon-
necting a
wireless
client.

Attacker's system

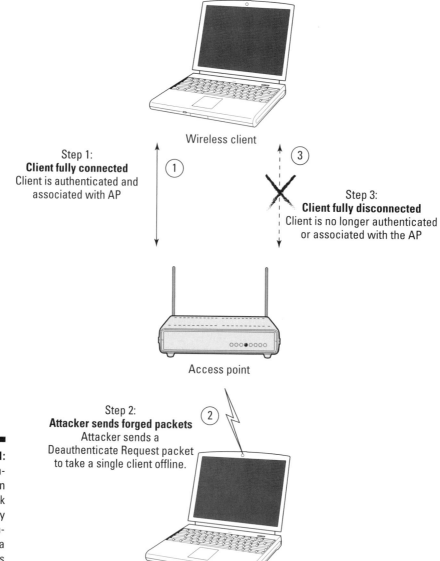

Wireless client

Step 1:
Client fully connected
Client is authenticated and
associated with AP

(1)

(3)

Step 3:
Client fully disconnected
Client is no longer authenticated
or associated with the AP

Access point

Step 2:
Attacker sends forged packets
Attacker sends a
Deauthenticate Request packet
to take a single client offline.

(2)

Figure 13-11:
Deauthen-
tication
attack
completely
discon-
necting a
wireless
client.

Attacker's system

If you care to see how your systems respond to deauthentication attacks, here's how it can be done using CommView for WiFi:

1. **Load CommView for WiFi and click the blue Start Capture icon in the upper-left corner or simply press Ctrl+S on your keyboard.**

   This loads the Scanner utility as shown in Figure 13-7 above so you can enable your wireless NIC to capture packets.

2. **Click the Capture button on the Scanner window.**

   This "opens" the Wireless Adapter Enable Promiscuous mode on your wireless NIC and allows you to start capturing wireless packets.

3. **Generate a Deauthentication packet.**

   It's a little trickier capturing one of these packets, but if you have an AP that supports manual deauthentications, capturing can be pretty simple. As shown in the Cisco management screen in Figure 13-12, it's as easy as clicking the Deauthenticate button for the client you wish to deauthenticate.

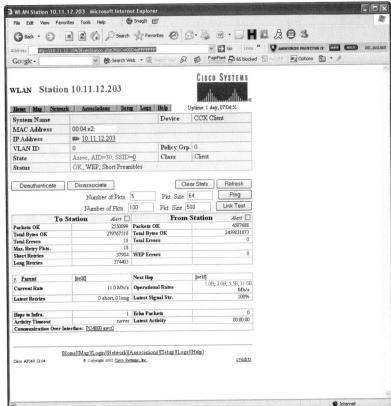

Figure 13-12: Cisco Aironet option to deauthenticate a wireless client.

4. Capture the Deauthentication packet.

This is as simple as capturing all wireless packets — or narrowing it down to management packets — in a network analyzer. Figure 13-13 shows what such a packet looks like in AiroPeek. All you have to do is capture the packet using any wireless network analyzer, save the packet, and import it into CommView for WiFi's Packet Generator. Or you can simply capture the packet in CommView for WiFi and save the packet using the steps we outlined for the Association Request packet above.

5. Edit the Deauthentication packet.

After you have the packet loaded into CommView for WiFi's Packet Generator, you can edit it to change source and destination addresses. In this example, we'll change the source address to effectively turn it into a forged address and change the destination address to the broadcast address.

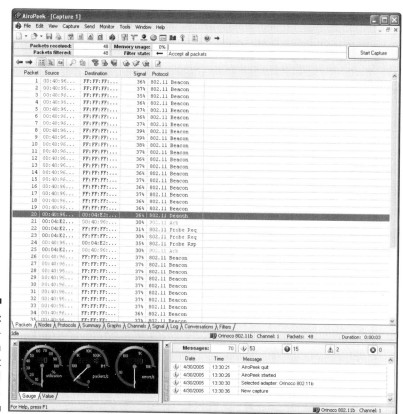

Figure 13-13: A Deauthentication packet discovered by AiroPeek.

Figure 13-14 shows the packet loaded into Packet Generator and edited to have a random source address (`11:22:33:44:55:66`) — and the broadcast address (`ff:ff:ff:ff:ff:ff`) as the destination address. You can change the BSSID address (MAC address of the AP) as well. These addresses and their locations within the packet are shown in Figure 13-14.

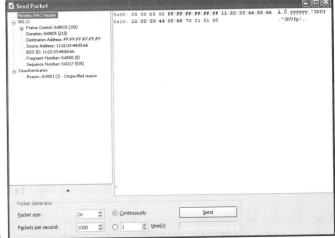

Figure 13-14:
An edited version of the Deau-thentication packet ready to send.

To edit the packet, you simply click inside the data area on the right side of the Packet Generator window and change the addresses to your heart's content. Just make sure you stay within the correct fields (offsets in hex editing terminology) so you don't overwrite other critical packet data. Note that in Figure 13-14, you can expand the 802.11 item on the left side (simply click the + button) and verify that your changes are accurate for the source, destination addresses, and even the BSSID address.

6. **Send the packet.**

   You can send the packet by setting the appropriate parameters for packet size, packets per second, and the number of times to send it.

This exercise demonstrates how simple it is to create a deauthentication flood attack against wireless clients. If you monitor your airwaves by a network analyzer (such as CommView for WiFi or AiroPeek) while you're performing this attack, you'll see quite a spectacle. Notice in Figure 13-15 how the majority of packets discovered by AiroPeek are Deauthentication packets.

Figure 13-16 shows what the same attack looks like through AiroPeek NX's Packets view. Notice that AiroPeek NX discovered the attack and highlighted the fact in the Expert column.

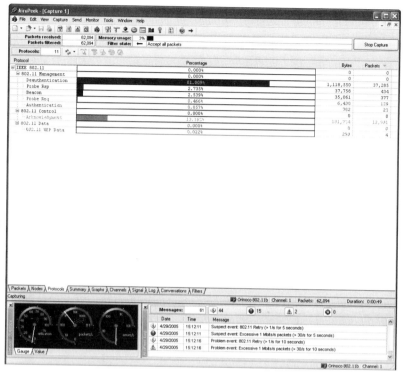

Figure 13-15:
Deauthen-
tication
attack as
seen in
AiroPeek's
Protocols
view.

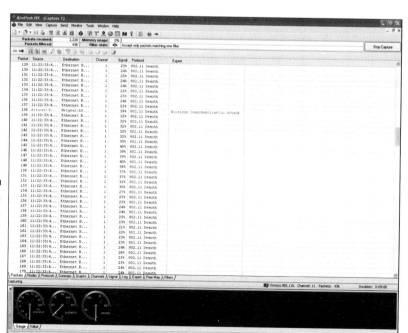

Figure 13-16:
Deauthen-
tication
attack as
seen in
AiroPeek
NX's
Packets
view.

For a real-world view of what this type of attack can do to a wireless client, take a gander at Figures 13-17 (normal wireless connectivity and a test ping out to a Web site) and 13-18 (the havoc after deauthentication).

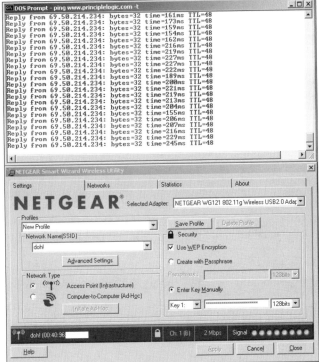

Figure 13-17:
Normal
wireless
client
connec-
tivity.

Invalid authentications via fata_jack

There are other tools that can create similar client DoS attacks. One popular one is Mark "Fat Bloke" Osborne's `fata_jack`. This is a Linux program based on the `wlan_jack` program that you'll have to compile before using. It sends out invalid Authentication Failed frames, allowing an attacker to spoof a valid client on the network and send these invalid frames to the AP. The AP, in effect, responds to the client with *Hey! Your previous authentication failed, so forget you — I don't want to speak to you any more.*

This attack is known to create erratic behavior on wireless clients, especially those running on older operating systems with older wireless hardware. Before using this program, you compile it (via the instructions in the source code); then you can run it to see whether any of your systems are vulnerable — just be careful so you don't crash critical systems.

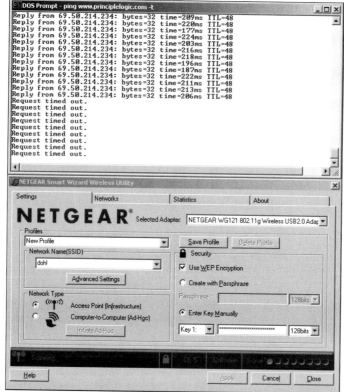

Figure 13-18: Wireless client connectivity losses after a deauthentication attack begins.

Physical Insecurities

When it comes to physical insecurities, we're not referring to that uncomfortable feeling when we realize we need to shed a few pounds. We're actually talking about an attacker physically exploiting an AP — maybe not with a sledgehammer, but with about that much subtlety — in a way that can lead to a DoS situation.

If an attacker wants to deny service to a reception area, a coffeeshop, or even an entire airport terminal, all he has to do is something trivial like shutting off the power or stealing the AP itself. If an attacker really wants to get sneaky, he could *slightly* unplug the Ethernet cable or *slightly* disconnect the antennae from the back of the AP. These two problems — easily and commonly overlooked — can drive you bonkers trying to troubleshoot!

When performing your ethical-hacking tests — or even if you're simply troubleshooting wireless-network connectivity problems — be sure to look at this oh-that's-too-obvious area. As do most people working in IT, we've found that the simple things tend to cause the most problems.

DoS Countermeasures

There are several things you can do to protect your airwaves and systems from DoS attacks. Many of these are free and relatively simple if you can spare the time. Only a couple of these suggestions require that you spend money — albeit *good* money — but the solutions are usually worth every penny.

Know what's normal

Establish a baseline of typical wireless-network usage. Use AiroPeek, CommView for WiFi, or your favorite network analyzer to look at

- ✔ Protocols in use
- ✔ Minimum, maximum, and average number of connections
- ✔ Minimum, maximum, and average throughput
- ✔ RF signal strength
- ✔ Any notable RF interference
- ✔ Number of users

It's best to gather this data as soon as you set up your network, if possible. If you can't do that, simply start now and use the data you gather as your baseline. Continue to monitor what's going on periodically, during

- ✔ Specific timeframes
- ✔ Random timeframes
- ✔ High-traffic times of day
- ✔ Low-traffic times of day

This information will prove invaluable when you're trying to determine whether a DoS attack is about to occur, is occurring, or has already occurred. Without baseline information, knowing what's right and what's not is maddeningly difficult.

Contain your radio waves

If RF signals are leaking outside your building — they likely are — then practically all of the DoS attacks we mention here are possible. Adding insult to injury, trying to track down where jamming signals are coming from (outside of using complex triangulation calculations) is very difficult.

The best way to keep your radio waves inside and intact is to use directional antennae whenever possible to point the signals in only the direction they need to go. You should also scale back the transmission power of your APs if possible. This can leave you more susceptible to stronger signals overpowering yours, but that's the chance you have to take. The Cisco Aironet AP shown in Figure 13-19 has this capability.

There's also RF shielding materials that can be built in to or added onto building walls and windows, but this can be costly. If actual shielding of the radio signals is not possible, then the best alternative is to keep attackers as far away from your wireless systems as you can. This means protecting your entire building — even your organization's campus — with fences and guard posts if necessary.

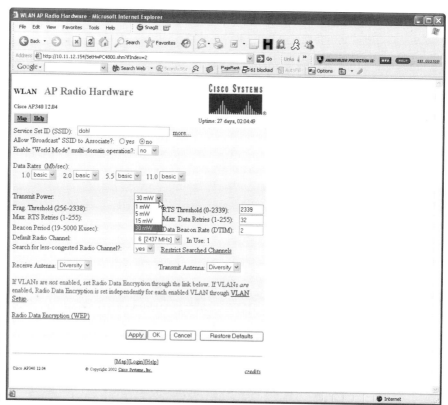

Figure 13-19:
Changing the transmit power on a Cisco AP.

If there's one good thing about Physical Layer jamming attacks, it's that hackers have to put forth a fair amount of effort to carry them out. If you keep upping the ante, they'll have to do that as well. Eventually they reach a limit: their jamming devices can only utilize so much transmitting power. If you put enough protective measures in place, the hackers will have to put themselves physically in the same room as your wireless systems — where they might be subject to not-so-subtle defensive measures (say, angry system administrators with two-by-fours). If you can't keep the bad guys out at this point, well, you may have bigger problems on your hands.

Limit bandwidth

Many enterprise-class APs allow you to tweak Quality of Service and Class of Service configurations to limit what comes and goes. Specific features vary among virtually every AP, so be sure to check your documentation to see what's available.

Use a Network Monitoring System

With a network monitoring system, you can set SNMP traps and other programmable alerts to notify you during excessive traffic loads, signal degradation, signal losses, and more.

Use a WIDS

A *wireless intrusion–detection/prevention system (IDS/IPS)* is perhaps the most effective way of defending against DoS attacks. Such systems look for

- Unauthorized MAC addresses
- Unauthorized broadcast traffic
- Jamming
- Association floods
- Authentication floods
- Disassociation attacks
- Deauthentication attacks

Most WIDS even track the state of wireless communications, and can look for various protocol anomalies. For instance, if data transfer is observed after deauthentication or disassociation requests, a WIDS system may smell a rat, determine that such requests are illegitimate, and tear down the communication link. Refer to Appendix A for a detailed listing of such vendors.

Attack back

Some WIDS already have the ability to attack the attacker, but you can do it yourself almost as easily. If bells, whistles, and automation are not in your budget, you can keep things simple by utilizing a tool such as Void11, CommView for WiFi, or other packet-injection programs, combined with a list of allowed systems on your network. If you come across an unauthorized system trying to attack your network, a simple deauthorization attack sent back in the attacker's direction may be all you need.

If your situation warrants fighting back, be very careful about it — you could end up breaking laws, violating security-ethics commandments, or simply getting schooled and trounced by your attacker.

Demand fixes

There are certain things that only the wireless-standards bodies (such as the IEEE and Wi-Fi Alliance) and the vendors of wireless products have control over. If you're serious about implementing wireless (and you or your organization have enough clout in the industry), then request — better yet, demand — that your wireless vendors and standards bodies fix the issues we cover in this chapter.

In addition, there's no reason organizations developing, testing, or certifying wireless-network products shouldn't use the same tools we demonstrate in this chapter and throughout this book. Again, in order to defend against the enemy, you must understand the enemy. Encouraging the powers that be to do so only makes logical sense.

Chapter 14

Cracking Encryption

Most people believe that encryption is a panacea. They believe that when you encrypt something, it's secure. Unfortunately, this is just not true. As with many newer technologies, you may find the available security features of encryption not as comprehensive or robust as you might like. Cryptography features can have flaws. You can use the wrong algorithm, a flawed algorithm, a short key, or a poor implementation, and (oops!) there it is: a security breach. This chapter demonstrates how one or more of these problems affects the use of encryption with your wireless networks.

But we don't want to play Cassandra and bring only bad news. We also show you some techniques for strengthening your access point. At a minimum, we strongly recommend that you use the built-in security features as part of an overall defense in-depth strategy.

What Can Happen

The IEEE 802.11 specification identified features that a wireless network needs to maintain a secure operating environment. One of the primary features was the use of encryption to provide the following:

✔ **Message privacy:** Sensitive information is encrypted when transmitted between two wireless entities to prevent interception and disclosure or prevent a third party from tracking communications between two other entities.

> ✔ **Message integrity:** An entity can verify that no one has changed the content of a message in transit.

Nice try. Even though the 802.11 standard attempted to address privacy and integrity, it fell well short. Let's look at these features and their shortcomings.

Protecting Message Privacy

The 802.11 standard supports privacy through the use of cryptographic techniques for the wireless interface. The first and most widely used algorithm was *Wired Equivalent Privacy,* also known as the *WEP algorithm.*

WEP uses the RC4 symmetric-key, stream-cipher algorithm to generate a pseudo-random data sequence. This *key stream* is simply added via a modulo 2 calculation (exclusive ORed) to the transmitted data. Unfortunately, all those syllables don't add up to very impressive security. (You can get a quick peek at why in the "Using Encryption" section, later in this chapter.)

Generally, the longer the encryption key, the harder it is to crack — but WEP (as defined in the 802.11 standard) supports only a puny (40-bit) size for the shared key. Fortunately, numerous vendors offered non-standard extensions of WEP that support key lengths from 40 bits to 104 bits. At least one vendor supports a key size of 128 bits (that is, 152 bits). The 104-bit WEP key, for instance, with a 24-bit Initialization Vector (IV) becomes a 128-bit RC4 key. WEP uses the IV to seed the algorithm before encrypting a frame.

In general — all other things being equal — increasing the key size increases the security of a cryptographic technique. But that isn't the whole story; flawed implementations or flawed designs can always prevent those long keys from increasing security. Research has shown that keys longer than 80 bits make *brute-force cryptanalysis* (running all possible key values on a superfast computer) a near impossible task — for robust designs and implementations, anyway. In practice, however, most WLAN deployments rely on the scrawny 40-bit keys specified in 802.11. And there's more bad news: Recent attacks have shown that the WEP approach to privacy is vulnerable to certain attacks *regardless of key size.* Brute-force attacks, mentioned above, are described later in the chapter.

Protecting Message Integrity

Making sure messages get through in tact is a basic security task. The IEEE 802.11 specification also outlined a simple *Cyclic Redundancy Check (CRC)* to provide data integrity for messages transmitted between wireless clients and access points. This security service was designed to reject any messages that anyone may have changed. The access point and client compute a CRC-32 or frame-check sequence called an *integrity check value (ICV)* for each frame prior to transmission. You can see in Figure 14-1 that WEP then encrypts the integrity-sealed packet, using the RC4 key stream to provide the ciphertext message. The receiver decrypts the frame and re-computes the CRC on the message. The receiving end then compares the computed CRC to the one computed with the original message. When the CRCs are not equal, an error occurs, and the receiver discards the frame.

Great idea, but (again) poorly implemented. It is possible to flip bits and still end up passing the CRC check. Bottom line: Message modification is possible, which makes CRC-32 inadequate for protecting against intentional data-integrity attacks. You need real cryptographic mechanisms — such as a secure hash, message digest, or message-authentication code (MAC) — to prevent deliberate attacks. Use of non-cryptographic mechanisms often facilitates attacks against the cryptography. In this case, it certainly does. One reason is the use of the 64- or 128-bit key for integrity *and* privacy, which is a cryptography no-no.

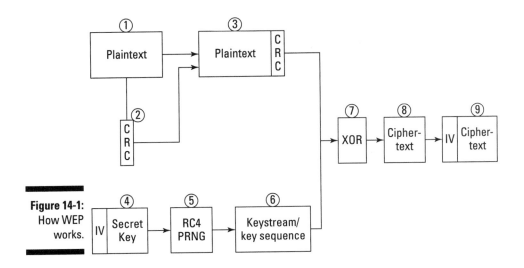

Figure 14-1: How WEP works.

Using Encryption

The popular press has done a lot to discourage organizations and individuals from using wireless networks. If you've been paying attention to the brouhaha, then you're aware of all the negative articles about wireless security — especially those dealing with encryption. Part of the problem is that the press and others don't understand the basis for WEP. As implied by its name, the developers of Wired Equivalent Privacy intended for it to give clients the same level of security found on a wired network — which, quite frankly, isn't much. With the exception of a fully switched environment, eavesdroppers can have their way with frames traversing a wired network. WEP was never intended to provide message integrity, non-repudiation, and confidentiality. And guess what — it doesn't.

WEP uses the symmetrical RC4 (Ron's Code 4) algorithm and a PRNG (Pseudo-Random Number Generator). The original standard specified 40 (in practice, 64) and 128-bit key lengths with a 24-bit initialization vector (IV). Then there's the matter of incomplete coverage of network layers: WEP encrypts Layers 3 through 7, but does not encrypt the MAC layer (that is, Layer 2). Because it's a symmetrical algorithm, WEP gives every client the keys and other configuration data.

Okay, we know there's nothing wrong with the RC4 algorithm per se — after all, Web browsers use it for Secure Sockets Layer (SSL). The problem is in the WEP implementation of the RC4 algorithm — and the false sense of security it encourages.

The algorithm takes the IV, which is in plaintext, and sticks it on the front end of the secret key (which the decrypter knows). WEP then plugs the result into the RC4 to regenerate the key stream. Next, the algorithm XORs the key stream with the ciphertext, which should give us the plaintext value. Finally, WEP re-performs the CRC-32 checksum on the message and ensures that it matches the integrity check value in our encrypted plaintext. Should the checksums not match, WEP assumes that someone tampered with the packet, and will discard it.

As mentioned earlier, access points generally have only three (namely, the following) encryption settings available:

- ✔ **None:** This setting represents the most serious risk because someone can easily intercept, read, and alter unencrypted data traversing the network.

- ✔ **40-bit shared key:** A 40-bit shared key encrypts the network communications data, but there is still a risk of compromise. The 40-bit encryption has been broken by brute force cryptanalysis, using a high-end graphics computer — and even low-end computers — so it has only questionable value. We show you some tools in later sections that allow you to easily recover 40-bit keys — and if you can, a bad guy can.

- ✔ **104-bit setting:** In general, 104-bit (sometimes called 128-bit) encryption is more secure than 40-bit encryption because of the significant difference in the size of the cryptographic key space. Even though this better security isn't true for 802.11 WEP (because of poor cryptographic design in the use of IVs), it is nonetheless recommended as a good practice. Again, you should be vigilant about checking with the vendor regarding upgrades to firmware and software — you may find some that overcome some of the WEP problems. (Some vendors, for example, support 152-bit keys.)

As a general rule, 40-bit keys are inadequate for any system. It is generally accepted that encryption keys should be longer than 80 bits to get the job done. The longer the key, the less likely that someone could compromise your access point by using a brute-force attack.

WEP Weaknesses

Security researchers have discovered security problems that let malicious users compromise the security of WLANs that use WEP — these, for instance:

- **Passive attacks to decrypt traffic:** These are based on statistical analysis.
- **Active attacks to inject new traffic from unauthorized mobile stations:** These are based on known plaintext.
- **Active attacks to decrypt traffic:** These are based on tricking the access point.
- **Dictionary-building attacks:** These are possible after analyzing enough traffic on a busy network.

 The biggest problem with WEP is when the installer doesn't enable it in the first place. Even bad security is generally better than no security.

When people do use WEP, they forget to change their keys periodically. Having many clients in a wireless network — potentially sharing the identical key for long periods of time — is a well-known security vulnerability. If you keep your key long enough, someone can grab all the frames he needs to crack it.

Can't blame most access-point administrators for not changing keys — after all, the WEP protocol doesn't offer any key management provisions. But the situation is dangerous: When someone in your organization loses a laptop for any reason, the key could become compromised — along with all the other computers sharing the key. So it's worth repeating . . .

 Shared keys can compromise a wireless network. As the number of people sharing the key grows, so does the security risk. A fundamental tenet of cryptography is that the security of a system is largely dependent on the secrecy of the keys. Expose the keys and you expose the text. Share the key, and a cracker only has to crack it once. Moreover, when every station uses the same key, an eavesdropper has ready access to a large amount of traffic for analytic attacks.

As if key management problems weren't enough, you have other problems with the WEP algorithm. Check out these bugbears in the WEP initialization vector:

- ✔ **The IV is too small and in cleartext.** It's a 24-bit field sent in the cleartext portion of a message. This 24-bit string, used to initialize the key stream generated by the RC4 algorithm, is a relatively small field when used for cryptographic purposes.

- ✔ **The IV is static.** Reuse of the same IV produces identical key streams for the protection of data, and because the IV is short, it guarantees that those streams will repeat after a relatively short time (between 5 and 7 hours) on a busy network.

- ✔ **The IV makes the key stream vulnerable.** The 802.11 standard does not specify how the IVs are set or changed, and individual wireless adapters from the same vendor may all generate the same IV sequences, or some wireless adapters may possibly use a constant IV. As a result, hackers can record network traffic, determine the key stream, and use it to decrypt the ciphertext.

- ✔ **The IV is a part of the RC4 encryption key.** The fact that an eavesdropper knows 24-bits of every packet key, combined with a weakness in the RC4 key schedule, leads to a successful analytic attack that recovers the key after intercepting and analyzing only a relatively small amount of traffic. Such an attack is so nearly a no-brainer that it's publicly available as an attack script and as open-source code.

- ✔ **WEP provides no cryptographic integrity protection.** However, the 802.11 MAC protocol uses a non-cryptographic Cyclic Redundancy Check (CRC) to check the integrity of packets, and acknowledges packets that have the correct checksum. The combination of non-cryptographic checksums with stream ciphers is dangerous — and often introduces vulnerabilities. The classic case? You guessed it: WEP.

There is an active attack that permits the attacker to decrypt any packet by systematically modifying the packet, and CRC sending it to the AP and noting whether the packet is acknowledged. These kinds of attacks are often subtle, and it is now considered risky to design encryption protocols that do not include cryptographic integrity protection, because of the possibility of interactions with other protocol levels that can give away information about ciphertext.

Only one of the problems listed above depends on a weakness in the cryptographic algorithm. Therefore substituting a stronger stream cipher will not help. For example, the vulnerability of the key stream is a consequence of a weakness in the implementation of the RC4 stream cipher — and that's exposed by a poorly designed protocol.

One flaw in the implementation of the RC4 cipher in WEP is the fact that the 802.11 protocol does not specify how to generate IVs. Remember that IVs are the 24-bit values that are pre-pended to the secret key and used in the RC4 cipher. The IV is transmitted in plaintext. The reason we have IVs is to ensure that the value used as a seed for the RC4 PRNG is always different.

RC4 is quite clear in its requirement that you should never, ever reuse a secret key. The problem with WEP is that there is no guidance on how to implement IVs.

Microsoft uses the RC4 stream cipher in Word and Excel — and makes the mistake of using the same keystream to encrypt two different documents. So you can break Word and Excel encryption by XORing the two ciphertext streams together to get the keystream to dropsout. Using the key stream, you can easily recover the two plaintexts by using letter-frequency analysis and other basic techniques. You'd think Microsoft would learn. But they made the same mistake in 1999 with the Windows NT Syskey.

The key, whether it's 64 or 128 bits, is a combination of a shared secret and the IV. The IV is a 24-bit binary number. Do we choose IV values randomly? Do we start at 0 and increment by 1? Or do we start at 16,777,215 and decrement by 1? Most implementations of WEP initialize hardware using an IV of 0; and increment by 1 for each packet sent. Because every packet requires a unique seed for RC4, you can see that at higher volumes, the entire 24-bit space can be used up in a matter of hours. Therefore we are forced to repeat IVs — and to violate RC4's cardinal rule against *ever* repeating keys. Ask Microsoft what happens when you do. Statistical analysis shows that all possible IVs (224) are exhausted in about 5 hours. Then the IV re-initializes, starting at 0, every 5 hours.

Other WEP Problems to Look For

As if the weaknesses in the algorithm weren't enough, other key vulnerabilities contribute to the problem. These vulnerabilities include WEP keys that are non-unique, never changing, unmodified factory defaults, or just bone-headed (weak keys made of all zeros or all ones, based on easily guessed passwords, or using other similarly trivial patterns).

One of the fundamental flaws of WEP is that it uses keys for more than one purpose. Generally, you don't use the same keys for authentication and encryption or the same key for integrity and privacy. Because WEP breaks these rules and others, it behooves you to protect your keys. Remember that WEP doesn't provide any help here. Break the authentication and you can break the encryption — and vice versa.

The manufacturer may provide one or more keys to enable shared-key authentication between the device that's trying to gain access to the network and the AP. And yes, we're going to say it again: Using a default shared-key setting is a security vulnerability — a common one because many vendors use identical shared keys in their factory settings. A malicious cracker may know the default shared key and use it to gain access to the network.

Don't use default WEP keys! No matter what your security level, your organization should change the shared key from its default setting because it's just too easily exploited. In the event you don't know the default keys for a wireless access point (or you don't know whether there is a default key), check out www.cirt.net.

Some products generate keys after a keystroke from a user that, when done properly using the appropriate random processes, can result in a strong WEP key. Other vendors, though, based WEP keys on passwords chosen by users; this typically reduces the effective key size.

You may find your configuration utility doesn't have a passcode generator, but allows you to enter the key as alphanumeric characters (that is, a to z, A to Z, and 0 to 9) rather than as a hexadecimal number. You just need to create a good passcode, right? Sounds like a good idea — until you study it. Each character you enter represents 8 bits, so you can type 5 characters for a 40-bit code and 13 characters for a 104-bit code. Entering 5 characters in ASCII is not as strong as generating the key randomly in hexadecimal. Think of all the poor five-letter passcodes you could create!

So take it from us: WEP is weak. The following is a summary of some of the more glaring weaknesses of WEP:

- The IV value is too short — and not protected against reuse.
- The way keys are constructed from the IV makes it susceptible to weak key attacks.
- There is no effective detection of message tampering; that is, WEP has no effective message integrity.
- It directly uses the master key and has no built-in provision to update the keys.
- There is no provision against message replay.
- There is no key-management mechanism built in.

At a minimum, enterprises should employ the built-in WEP encryption. But that's a poor minimum. And it's amazing how many access points don't have any encryption at all. We find that less than half the access points we stumble on have encryption of any sort.

If an access point is using WEP, it makes your hacking a *little* more difficult but certainly not impossible. You just need to get yourself a WEP cracker. Several are available from Web sites on the Internet and are relatively easy to use. If you've dug in to Chapter 8 and know how to use `ethereal`, then cracking WEP keys is easy.

How long it will take to crack the WEP key depends on the access point's level of activity.

Attacking WEP

There are several active and passive attacks for WEP as follows:

- ✔ Active attacks to inject traffic based on known plaintext
- ✔ Active attacks to decrypt traffic based on tricking access points
- ✔ Dictionary-based attacks after gathering enough traffic
- ✔ Passive attacks to decrypt traffic using statistical analysis

The following sections discuss these attacks in detail.

Active traffic injection

Suppose an attacker knows the exact plaintext version of one encrypted message using a passive technique. The attacker can use this information to construct — and insert — correctly encrypted packets for the network. To do this, the attacker constructs a new message calculating CRC-32 values and performs bit-flips on the original message to encrypt plaintext in its encrypted form. The attacker can now send the packet to the access point, undetected. There are several variations on this technique; here's where you get the tools to use them:

- ✔ **Aireplay:** This program lets you take any captured packet and reinject it back onto the network.
- ✔ **WEPWedgie** (`http://sourceforge.net/projects/wepwedgie/`): This program is a toolkit for determining 802.11 WEP keystreams and injecting traffic with known keystreams.

Active attack from both sides

An extension of the active injection technique from the previous section. The attacker makes guesses on packet header contents rather than packet payload. Bit-flipping can transform destination addresses and route traffic to rogue devices where retransmission (with alterations) could occur. Educated guessing can also provide port information to allow passage through firewalls by changing it to use port 80 (the default port for Web traffic).

Table-based attack

A small space of possible WEP initialization vectors (IVs) — and the high likelihood they'll be reused at relatively short intervals — allow attackers to build decryption tables. Using passive techniques, the attacker gains some plaintext information. The attacker can then compute the RC4 key stream used by the IV. Over time, repetitive techniques allow an attacker to build a complete decryption table of all possible IVs. This allows an attacker to decipher every packet sent.

Passive attack decryption

This is more of an intrusion than an attack, but monitoring leads to further exploits. An attacker monitors traffic until an IV *collision* occurs. A collision is when the algorithm reuses an IV. When a collision happens, the shared secret and the repeated IV results in a key stream that has been used before. Because the algorithm sends the IV in ciphertext, an attacker keeping track of all the traffic can identify when collisions occur. Then, the attacker will use the resulting XOR information to infer data about the message content.

IP traffic is redundant in nature, and replication of this process easily yields enough data to decipher the encrypted text.

You can find commercial, off-the-shelf (COTS) hardware readily available to monitor 2.4 GHz transmissions. We cover some of these products in Chapter 8. By re-configuring drivers, you can cause the hardware to intercept encrypted traffic. Using the techniques described previously, you can make the WLAN vulnerable.

Cracking Keys

We have discussed a lot of WEP flaws in this chapter, with good reason: WEP is the algorithm most commonly used to protect wireless networks. But WEP has many flaws. These flaws leave WEP open to crack attacks. To crack WEP keys, you need

- A large amount of captured frames
- A program to process the frames

. . . and that's about all. Then you use the tools identified in Chapters 8 and 10 to capture frames for you. Okay, they don't crack the keys for you, but that's not much of a problem: You simply use another tool, such as WEPcrack or AirSnort. To add to your store of goodies, you can have a program like Kismet save weak IVs to feed into another program such as WEPcrack.

Using WEPcrack

WEPcrack (`http://sourceforge.net/projects/wepcrack/`) is perhaps the most famous of all WEP crackers. Most likely WEPcrack made its reputation as it was the first tool to hit the street. WEPcrack captures, logs, and cracks IVs to provide keys.

All you need to run WEPcrack is some packets and PERL. The WEPcrack authors wrote it so it is portable anywhere there is a PERL interpreter. That's easy to meet for most UNIX platforms. It's simple to run WEPcrack in UNIX, just type `perl /tmp/WEPcrack.pl` at the prompt (assuming that's where you installed the script). Running WEPcrack is conceivably a challenge for Windows users because Microsoft does not provide PERL natively. But you can use Cygwin (if you didn't install it already, you might want to refer to Chapter 4) or you can get yourself a PERL interpreter for Windows, such as ActivePerl.

Should you have a Windows platform, you'll need to download and install ActivePerl. You can download the freeware ActivePerl from ActiveState (`www.activestate.com/Products/ActivePerl/`). Installing and using ActivePerl to run WEPcrack is as easy as following these steps:

1. **Start ActivePerl setup.**

   You should see the setup window as shown in Figure 14-2.

2. **Click Next.**

   The license agreement appears on-screen.

3. **Select the** I accept the terms in the License Agreement **radio button and then click Next.**

   Another window appears.

4. **Choose where you want to install ActivePerl and then click Next.**

   If you don't want to install it at the root directory, click the Browse button and browse your directory tree until you find the location where you want to install. If you click Browse, you see the window shown in Figure 14-3.

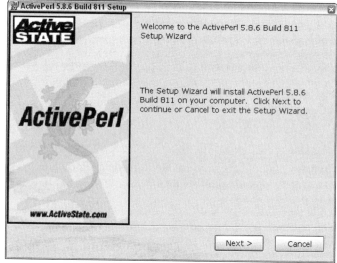

Figure 14-2:
The
ActivePerl
installation
wizard.

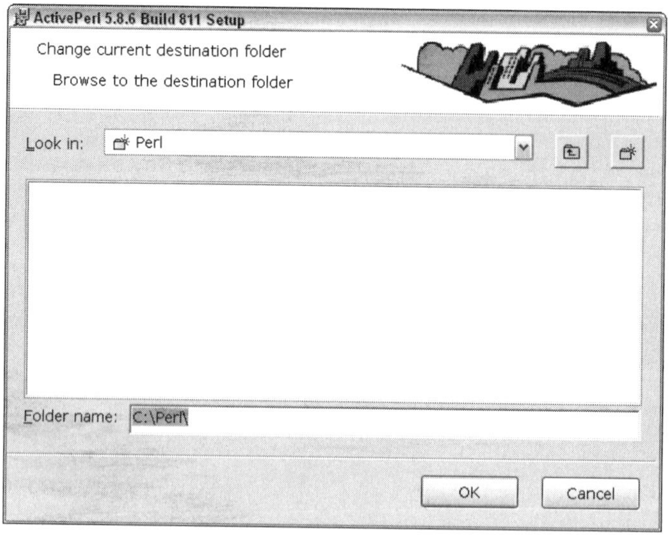

Figure 14-3:
Custom
setup
window.

5. **In the New features in PPM window, either select Enable PPM3 to send profile info to ASPN; otherwise click Next.**

6. **If you wish, deselect any options you won't be using and then click Next.**

7. **Click the Install button.**

   The Install Status window appears. As it says in the window, installation can take several minutes. When the process is complete, you are ready to use PERL.

8. **From the Start menu, select Run.**

9. **Type** `command` **in the Open: dialog box and then click OK.**

10. **At the command prompt, type** `perl \progra~1\wepcrack\pcap-getIV.pl`.

    Of course, you'll have to substitute the correct path for your WEPcrack PERL routines; the one we show here is just an example. This script itself is useless unless you have a captured packet you can drop in from another program such as `ethereal` or `prismdump`. You will need about 5 million frames; then you can proceed to Step 11.

    If you have `prismdump` and want to do it in one step, you can run the command `prismdump | pcap-getIV.pl`. The output of this step is the `ivfile.log` file.

11. **Start WEPcrack by typing** `perl \progra~1\wepcrack\wepcrack.pl ivfile.log`.

    Oops, there goes another WEP key.

12. **When you are finished, click the X in the upper right-hand corner of the window.**

    Unfortunately, the key is in decimal format, so you have to convert it to hexadecimal before you can use it.

Using AirSnort

AirSnort (`http://airsnort.shmoo.com/`) is a passive scanner that also cracks WEP keys. When AirSnort gathers enough weak initialization vectors, it starts to crack the WEP key. Of the over 16 million IVs, approximately nine thousand of the 128-bit keys are weak. AirSnort looks for these weak keys. The folks at the Shmoo Group estimate they need only about 2,000 weak IVs to guess the WEP key.

You can download AirSnort from Sourceforge at

`http://sourceforge.net/projects/airsnort/`

It runs on a UNIX platform, and is fairly easy to install and use. Windows users can also get WinAirSnort from Nevillon at

`www.nwp.nevillon.org/attack.html`

Figure 14-4 shows the WinAirSnort window that looks eerily like the Linux version.

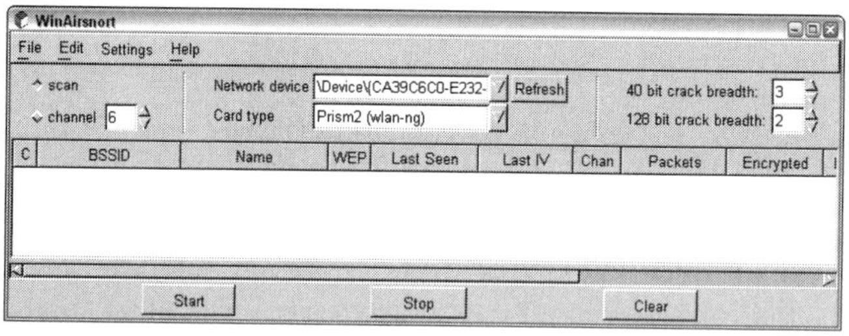

Figure 14-4:
WinAirSnort
window.

To install AirSnort in Linux, follow these steps:

1. **Unzip the download. To do so, use the following command:**

   ```
   /#gunzip airsnort-0.2.7e.tar.gz
   ```

2. **Untar the file with the following command:**

   ```
   /#tar -xvf airsnort-0.2.7e.tar
   ```

3. **Change the directory to the one you created when you uncompressed and retrieved the archive. Do this with this command:**

   ```
   /#cd airsnort-0.2.7e
   ```

4. **Compile and install AirSnort.**

   You may find your platform requires different commands, but you get the idea. The compiling procedure puts AirSnort binaries in the /user/ local/bin directory. Depending on your platform, try one of these commands:

   ```
   /airsnort-0.2.7e# ./autogen.sh
   /airsnort-0.2.7e# make
   /airsnort-0.2.7e# make install
   ```

5. **To run AirSnort, open a terminal window and type the following command:**

   ```
   /airsnort-0.2.7e# airsnort
   ```

6. **Use the up or down arrows to select the channel you want to scan.**

   If you used Kismet or another wireless scanner earlier to identify channels, then you most likely know the channel you want to monitor. If you simply want to monitor all the channels, click Scan.

7. **From the** Network device **drop-down list, select your network device.**

   This is the device that you will use to monitor. For example, select eth0. AirSnort does not necessarily put cards into monitor mode automatically.

8. **From the Card type drop-down list, select your wireless NIC.**

   For example, you can select ORiNOCO.

9. **If you want to decrease the time it takes to crack the key, then increase the 40-bit or 128-bit crack breadth.**

   Increasing the *crack breadth* increases the number of key possibilities examined when AirSnort attempts to break the WEP key.

10. **Click the Start button on the lower-left part of the bottom bar.**

    AirSnort will start to show you some interesting SSIDs — and eventually will crack the key.

It can take a long time to crack the key. The Shmoo crew gives an example in its FAQ (`http://airsnort.shmoo.com/faq.html`). Take the example of a small company with four employees. The four use the Internet all day and in so doing generate about 1 million packets per day. Of those 1 million packets, about 120 are interesting. Regardless of whether you look at the total packets or the interesting IVs, the keys are exhausted in about 16 days. The more employees, the shorter the timeframe.

You'll find a wealth of information at the Shmoo site including WEP implementation and passive monitoring.

Using aircrack

Aircrack (`www.cr0.net:8040/code/network/`) from Christophe Devine is another WEP-cracking tool. There is a Windows and Linux version. However, we will concentrate in this section on the version made specifically for the Windows platform.

Aircrack implements KoreK's attacks as well as improved FMS (Fluhrer-Mantin-Shamir) attacks. Aircrack provides the fastest and most effective statistical attacks available. To give aircrack a try, simply collect as many packets as possible from a WEP-encrypted wireless network, and then start aircrack. Perhaps a few steps to illustrate this:

1. **Download and unzip** `aircrack-2.1.zip` **from Christophe's Web site.**

   After you unzip the `aircrack-2.1.zip` file, you should see in the `win32` subdirectory a file titled `airodump`. Airodump is a packet-capture program.

2. **Start the packet capture by double-clicking the airodump icon.**

   You should see the window shown in Figure 14-5.

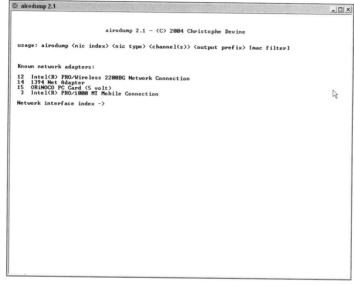

Figure 14-5:
Airodump
window:
starting the
options.

3. **From this list of known wireless interfaces, pick the one you want to use, choose it, and hit the Enter key.**

Your window should look similar to the one shown in Figure 14-6.

4. **Choose your interface and then press Enter.**

You can choose **o** for Orinoco/Realtek interfaces or **a** for Aironet/Atheros interfaces.

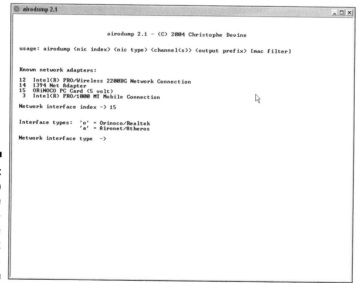

Figure 14-6:
Airodump
window: the
network-
interface
index
option.

5. **Enter a channel to scan and then press Enter.**

   If you know the channel from your wardriving, then enter the number of the channel you want to dump. Otherwise enter **0** (zero) to scan them all.

6. **Enter any name for the output file, and then hit the Enter key.**

   Pick a name that makes sense. You may want to include the date and time in the name.

7. **If you want to filter on a particular MAC address, enter it. Otherwise type p for none. Hit the Enter key.**

   Your window should look similar to the one shown in Figure 14-7.

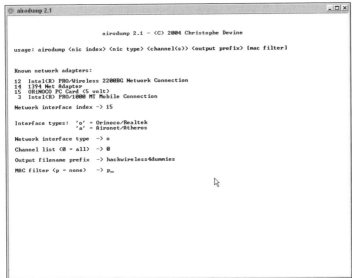

Figure 14-7:
Airodump
window:
finishing the
options.

8. **Observe as Airodump starts capturing frames.**

   A window like the one in Figure 14-8 appears. From this window, you can see airodump racking up the IVs.

9. **Double-click the aircrack icon.**

   A window like the one in Figure 14-9 appears.

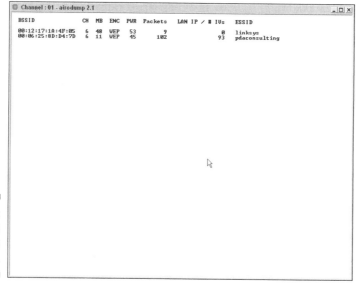

Figure 14-8:
Airodump
capturing
frames.

Aircrack often determines a WEP key within a few seconds, but the execution time is highly variable. It took Peter several days to crack a WEP with little activity and a 40-bit key. Shorter execution times require more traffic, more unique IVs, more luck, and the lowest successful *fudge factor,* a setting that tells aircrack how wildly it should guess when trying new keys. The higher the fudge factor, the more keys aircrack will try — increasing both the potential time of execution and the likelihood that the attack will succeed. The fudge factor has a default value of two, but you can set it to any positive integer. The default setting is a good place to start, but try several different settings when the initial attack does not succeed. Note, however, that there's a tradeoff: Generally the higher the fudge factor, the longer the execution time.

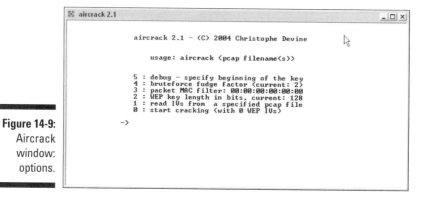

Figure 14-9:
Aircrack
window:
options.

10. **If you want to use a fudge factor other than the default, type it in and then press Enter.**

11. **Access the file created in Steps 1 through 9.**

    You can type the name and hit the Enter key or (following aircrack's suggestion) drag the file over to where you need it.

12. **Enter a** 0 **(zero) and then hit the Enter key.**

    If you get the message shown in Figure 14-10, then you need to let airo-dump gather more IVs. Should you see the message in Figure 14-11, however, it's Game Over time.

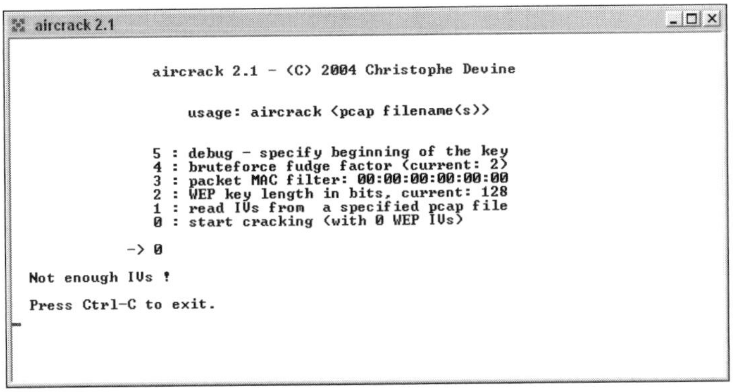

Figure 14-10: The Aircrack IV message.

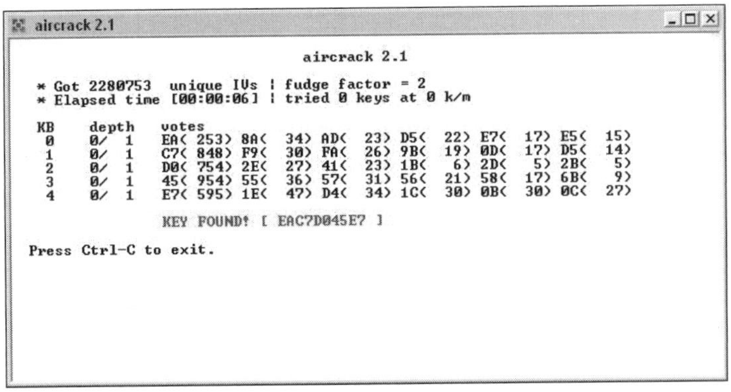

```
aircrack 2.1 - (C) 2004 Christophe Devine

        usage: aircrack <pcap filename(s)>

        5 : debug - specify beginning of the key
        4 : bruteforce fudge factor (current: 2)
        3 : packet MAC filter: 00:00:00:00:00:00
        2 : WEP key length in bits, current: 128
        1 : read IVs from  a specified pcap file
        0 : start cracking (with 0 WEP IVs)

        -> 0

Not enough IVs !

Press Ctrl-C to exit.
```

Figure 14-11: Aircrack success!

```
                    aircrack 2.1

 * Got 2280753  unique IVs ; fudge factor = 2
 * Elapsed time [00:00:06] ; tried 0 keys at 0 k/m

KB    depth     votes
 0    0/  1     EA( 253) 8A(  34) AD(  23) D5(  22) E7(  17) E5(  15)
 1    0/  1     C7( 848) F9(  30) FA(  26) 9B(  19) 0D(  17) D5(  14)
 2    0/  1     D0( 754) 2E(  27) 41(  23) 1B(   6) 2D(   5) 2B(   5)
 3    0/  1     45( 954) 55(  36) 57(  31) 56(  21) 58(  17) 6B(   9)
 4    0/  1     E7( 595) 1E(  47) D4(  34) 1C(  30) 0B(  30) 0C(  27)

            KEY FOUND! [ EAC7D045E7 ]

Press Ctrl-C to exit.
```

Using WepLab

WepLab (http://weplab.sourceforge.net/) from Jose Ignacio Sanchez provides an alternative implementation of the KoreK attacks. Similar to aircrack's fudge factor, WepLab provides a probability adjustment with its percent (--perc) command-line option. The default setting of 50 percent is fairly aggressive, and results in relatively few branches; higher settings increase the number of branches taken. In addition to excellent statistical attacks, WepLab provides brute-force and dictionary-cracking attacks that can be very effective. This combination of techniques makes WepLab an essential tool.

Finding other tools

Don't like any of those tools? Don't fret — there are more. For example, you can try these WEP crackers:

- ✔ Auditor Security Collection: `http://remote-exploit.org/?page=auditor`

- ✔ chopchop: `www.netstumbler.org/showthread.php?t=12489`

- ✔ Dwepcrack: `www.e.kth.se/~pvz/wifi/`

- ✔ jc-wepcracker: `www.astalavista.com/?section=dir&cmd=file&id=3316`

- ✔ WepAttack: `http://wepattack.sourceforge.net/`

Not sure what one to use? Well, we don't usually enter into religious debates. But we can point you to a place where they do. You can find an excellent comparison of the above tools at SourceForge (`www.securityfocus.com/infocus/1814`).

Armed with a WEP key, you can decipher new packets you gather using `AirSnort`, `ethereal`, or `prismdump`. Or you could use `WEP Decrypter` (`www.linkferret.ws/products/decrypt.htm`), `wep_decrypt` (`www.lava.net/~newsham/wlan/wep_tools.tgz`), or `chopchop` (`http://forums.netstumbler.com/showthread.php?t=12489`) to decrypt frames. You also can negotiate with the access point and gain entry to the network. Once on the network, you can run `nmap` (`www.insecure.org/nmap/`) or `Cain & Abel` (`www.oxid.it/cain.html`) to gather passwords, discern applications, and fingerprint operating systems.

Countermeasures Against Home Network-Encryption Attacks

If you've read up to this point, we wouldn't blame you if you're a bit crestfallen about using wireless networks safely. Don't despair, there are some things you can do to protect yourself — even at home. You can

✔ Rotate the keys

✔ Use Wi-Fi Protected Access (WPA)

These are discussed in detail in the sections that follow.

Rotating keys

As mentioned, WEP is a *symmetric algorithm* that uses the same secret key for encryption and decryption. Sender and receiver must possess the same key. If they must share keys, then they must also have a method for securely exchanging those shared keys.

Amazingly, there is nothing in the 802.11 standard about managing keys — even though key management is probably the most critical aspect of a cryptographic system. But key management for 802.11 systems largely left up to the users of a 802.11 network — many of whom you wouldn't exactly call security-savvy. Result: Many vulnerabilities are introduced into the WLAN environment. The lack of key management in the original 802.11 specification also leaves key distribution unresolved. Without a consistent sense of how to exchange keys securely, WEP-secured WLANs don't scale well.

When an enterprise recognizes the need to change keys often and to make them random, the task is formidable in a large WLAN environment. When you have only two or three laptops, this is an annoyance; when you have 3,000 workstations, it's a potential showstopper. Each one of those 3,000 workstations must have the *same* secret key — *and* the owner of every workstation must keep it secret. Generating, distributing, loading, and managing keys for an environment of this size is a significant challenge and well-nigh impossible. Compromise one client and you have the keys.

You know what they say about secrets? (Here's a hint: It's no secret.) Have you ever lost a laptop? Have you ever lost an employee? In both cases, you should change *all 3,000 keys.* Otherwise someone can decrypt every message, because everybody is using the same key. And just how often do you *really* think administrators will change the keys?

Using WPA

You may have heard of 802.11i, the new kid on the standards block. Check out the IEEE (`www.ieee.org/`), if you haven't. IEEE 802.11i defines the "robust security network (RSN)." An access point that meets this standard will only allow RSN-capable devices to connect. RSN is the environment we are evolving to because it provides the security services we require for a network.

Only time will tell whether there are flaws in 802.11i. We cover 802.11i features in this section — and again, later in the chapter, when we cover AES. Implementing 802.11i requires new hardware. Not everyone wants (or needs) to acquire new hardware — but everybody will still want improved security. So far, it looks as if WPA comes to the rescue.

An initiative for improving WLAN security is the interim solution — Wi-Fi Protected Access (WPA) — to address the problems of WEP. WPA uses the Temporal Key Integrity Protocol (TKIP) to address the problems without requiring hardware changes — that is, requiring only changes to firmware and software drivers. TKIP is also part of the RSN.

WPA is an example of a software or firmware patch. As an interim security solution, WPA does not require a hardware upgrade to your existing 802.11 equipment; the full-blown 802.11i standard does. WPA is not a perfect solution, but it does attempt quick and proactive delivery of enhanced protection to address some of the chronic WEP problems that predate the availability of 802.11i security features. WPA has two key features:

- 802.1X support
- Temporal Key Integrity Protocol (TKIP)

WPA uses 802.1X port-access control to distribute per-session keys. (Some vendors previously offered 802.11X support, even though it wasn't specified in the standard.) The 802.1X port-based access control provides a framework that allows use of robust upper-layer authentication protocols.

Temporal Key Integrity Protocol (TKIP) provides key mixing and a longer initialization vector. It also provides a Message Integrity Check (MIC) that prevents wireless data from being modified in transit. Even better, TKIP offers some essential support for wireless-network security by

- Managing keys to prevent the reuse of a static key
- Facilitating the use of session keys because cryptographic keys should change often
- Including four new algorithms to enhance the security of 802.11
- Extending the IV space
- Allowing for per-packet key construction
- Providing cryptographic integrity
- Providing key derivation and distribution

Through 802.11i and WPA, TKIP protects against various security attacks discussed earlier in this chapter — including replay attacks and attacks on data integrity. Additionally, it addresses the critical need to change keys. Again, the objective of WPA was to bring a standards-based security solution to the marketplace to replace WEP until full-blown IEEE 802.11i Robust Security Networks (RSNs), based on an amendment to the existing wireless LAN standard, become available. RSN also includes the Advanced Encryption Standard (AES) for confidentiality and integrity.

But WPA is not without its problems. Basically, one can crack Wi-Fi Protected Access Pre-Shared Keys that use short passphrases based on words found in the dictionary (yes, people still do that). For WPA, certain short or dictionary-based keys are easy to crack because an attacker can monitor a short transaction — or force that transaction to occur and then perform the crack remotely.

You will find software to help with WPA cracking as well. The `WPA Cracker` (`www.tinypeap.com/page8.html`) tool is somewhat primitive, requiring that you enter the appropriate data retrieved via a sniffer. (Your friendly authors recommend that you use `ethereal`.)

So how do you protect yourself when using WPA? Well, you can

- ✓ Choose a better passphrase, especially one that isn't made up of words in the dictionary.
- ✓ Select a passphrase that is at least 20 characters long.
- ✓ Randomly choose your passphrase.
- ✓ Use WPA Enterprise or 802.1X with WPA.
- ✓ Use virtual-private-network (VPN) technology, such as those technologies described later in this chapter.

As with all encryption matters, RSN, WPA, and TKIP are fairly complex topics. You can find more information on these protocols and algorithms in Peter's *Wireless Networks For Dummies* (Wiley).

Organization Encryption Attack Countermeasures

Your organization should adopt the techniques provided for the home network where practical. But you should supplement those features with one of these measures:

- ✓ WPA2 technology
- ✓ VPN technology

These are discussed in detail in the following sections.

Using WPA2

As much of an improvement as it is, WPA is still based on the RC4 algorithm — a stream cipher. But a major component of RSN is the use of the Advanced Encryption Standard (AES) for both data confidentiality and integrity. Presently, you can find AES WRAP (Wireless Robust Authenticated Protocol) products, but the final specification requires the AES-CCMP (Counter Mode-Cipher Block Chaining MAC Protocol) algorithm.

WPA2 (as specified in 802.11i) helps prevent replays and repudiation, enhances message integrity, and protects message privacy.

The 802.11i specification offers Advanced Encryption Standard (AES)-based cryptographic services at the Data Link Level and validates them according to the federal standard FIPS 140-2. Because AES will mitigate most concerns you may have about wireless eavesdropping or active wireless attacks, we strongly recommend its use. Keep in mind, however, that a wireless protocol at the Data Link Level protects only the wireless *sub*network — not the *entire* network. Where traffic traverses other network segments — whether those are local- or wide-area networks, wired segments, the Internet, or your in-house network's backbone, you may require additional security. That means implementing higher-level, FIPS-validated, end-to-end cryptographic protection.

The AES-based solution provides a highly robust security stance for the future — but requires new hardware and protocol changes. Your organization may have difficulty justifying the use of AES because it requires you to build a Public Key Infrastructure (PKI) — and that's costly.

At the time of publication, we could not find any cracking tools for AES-CCMP. That doesn't mean they don't exist or won't emerge; it just means you get a reprieve for now. But you are in an arms-escalation race. The crackers will catch up, so you must remain ever vigilant.

Using a VPN

Your organization can supplement the other controls in this book with a *virtual private network (VPN)* — a network that is created using public wires to connect private nodes. It's essentially a secure "tunnel" through the Internet; its "walls" are made of high-level encryption measures. It's attractive because it normally means less investment in hardware; many of us, in fact, are already using the Internet to connect to office applications. But the Internet is a very public network — and the public is partly made up of bad guys.

Even accessing your e-mail from the park outside your office as you sip your latté is risky without a VPN — because the person next to you on the bench could use the tools in this chapter to intercept and decrypt all your work. If you make frequent use of a wireless network at the office, you need to install and use VPN technology to protect yourself.

There are three general types of VPN:

- ✔ **Remote Access VPN:** This, the most common VPN, allows a remote user to securely access internal applications such as e-mail.

- ✔ **Extranet VPN:** This allows one organization to securely access another organization.

- ✔ **Intranet VPN:** In this VPN, data crossing the organization's normal network is encrypted.

You will find many solutions that help you to create networks using the Internet as the medium for transporting data. Typically, VPN solutions use encryption to ensure that only authorized users can access the network and that nobody can intercept the data. The solutions provide a tunnel between two networks that only authorized persons can access. You set up a tunnel each time you need it, and it is torn down when you are finished. In lieu of end-to-end cryptographic applications, your organization may find that it is necessary to build tunnels over public networks at the network or transport layer. There are many VPN solutions, ranging from commercial applications to sophisticated features that are available as part of our operating systems. Some of the more popular protocols for VPNs are:

- ✔ Point-to-Point Tunneling Protocol (PPTP)
- ✔ Layer 2 Tunneling Protocol (L2TP)
- ✔ Internet Protocol Security (IPSec)
- ✔ Secure Shell (SSH)

These are discussed in detail in the following sections.

Using Microsoft's Point-to-Point Tunneling Protocol

Arguably the weakest of all techniques, PPTP offers a quick and relatively painless method of accessing your network. It does offer a level of encryption that is more than adequate for most small-business owners. Small-to-medium-size businesses tend to like PPTP because it doesn't need a certificate server (as do L2TP and IPSec) and it supports native Windows commands. Finally, client software is available for all Microsoft operating systems and most commercial VPN vendors support PPTP.

You can choose authentication that uses passwords. Understand, however, that PPTP relies heavily on your password-generation skills.

Using Layer 2 Tunneling Protocol

Microsoft has made Layer 2 Tunneling Protocol (L2TP) available on the Windows 2000 or 2003 platform. Its primary drawback for the small business owner is its need for a certificate server or third-party certificate — which may not be affordable.

Using IPSec

IPSec is an industry standard for encryption that Microsoft includes in its newer Windows 2000, XP and 2003 operating systems. It is reasonably easy to set up between Windows machines and offers excellent security. Like L2TP, IPSec requires the use of a certificate server or a third-party certificate.

IPSec has two modes of use: tunnel and transport. Tunnel mode encrypts the header and the payload of each packet, while transport mode only encrypts the payload.

A rule of usage: Tunnel on the WAN and transport on the LAN.

Using SSH2

SSH or Secure Shell is another tunnel. Organizations commonly used SSH to tunnel services with cleartext passwords such as Telnet and FTP.

SSH also allows you to log in to remote host computers securely — as we did earlier using PPTP. You can also run commands on a remote machine, and enjoy secure, encrypted, and authenticated communications between two machines or networks. Within this tunnel, you run the services you want to protect such as e-mail, FTP, or even Web browsing.

We don't mean to leave you with the impression that a VPN is a silver bullet. The bad guys mount client-side and server-side attacks on VPNs. If your weakness is your clients, then that's where they attack. If you are interested in more information on WLANs and VPNs, pick up *Wireless Networks For Dummies* (Wiley).

Chapter 15

Authenticating Users

*I*n this chapter, we provide an overview of the user-authentication mechanisms to better illustrate their existing limitations. By *authentication,* we mean that one entity has to prove its identity to the other before a secure transaction can take place. Because 802.11-based wireless LANs are relatively new technologies, you may find the available authentication features are not as comprehensive or robust as you would like. We'd agree. Fortunately, each new standard improves upon existing standards. But it's an ongoing battle; we can always find tools to exploit weaknesses described in this chapter. Those flaws arise partly because the vendors frequently disable built-in security features as default settings — and partly because people either fail to change the defaults, or don't use built-in security features as part of their overall defense in-depth strategy.

Three States of Authentication

A basic, indispensable security service is authentication. In the standard 802.11, we don't authenticate users, but we authenticate machines. Under some common (but vulnerable) security setups, if you want people to authenticate their machines to a particular access point, then you make sure they know the shared key. (Bad idea.) This chapter shows you why you don't want to use the shared key to authenticate.

First off, it's worth looking at the three states a wireless client goes through in the authentication process:

- ✔ **Unauthenticated and unassociated:** The client selects a basic service set by sending a probe request to an access point with a matching SSID.

- ✔ **Authenticated and unassociated:** The client and the access point perform authentication by exchanging several management frames. Once authenticated, the client moves into this state.

- ✔ **Authenticated and associated:** Client must send an association request frame, and the access point must respond with an association response frame.

A client can authenticate to many access points, but will associate only with the access point with the strongest signal.

In the second state, we just casually mention the client authenticates to the access point. It's not quite that simple.

Authentication according to IEEE 802.11

The IEEE 802.11a and b specifications define two ways to "validate" wireless users who are attempting to gain access to a wired network. One does the job; the other one doesn't:

Open-system authentication: convenient but dangerous

This "authentication" technique isn't really authentication because the access point accepts the mobile station willy-nilly without verifying its identity. The access point authenticates a client when the client simply responds with a MAC address during the two-message exchange, in a simple (and insecure) process:

1. Client makes a request to associate to an access point.

2. The AP authenticates client and sends a positive response — voilà! The client is associated.

Shared-key authentication

Shared-key is a cryptographic — that is, real — technique for authentication. It is a simple "challenge-response" scheme based on whether a client has knowledge of a shared secret. In this scheme, the access point generates a random 128-bit challenge that it sends to the wireless client. The client, using a cryptographic key that is shared with the access point, encrypts the challenge or *nonce* (as it is called in security vernacular) and returns the result to the access point. Then, the access point decrypts the result that the client computed and sent, and allows access only when the decrypted value is the same as the random challenge transmitted. The algorithm used in the cryptographic computation and for the generation of the 128-bit challenge text is the same RC4 stream cipher used for encryption.

This authentication method is a rudimentary cryptographic technique at best; it doesn't provide mutual authentication. That is, the client does not authenticate the access point — therefore there's no assurance that a client is communicating with a legitimate wireless network. It is also worth noting that simple unilateral challenge-response schemes have long been known to be weak. They suffer from numerous attacks, including the infamous "monkey-in-the-middle" attacks covered in Chapter 12.

At least the shared-key authentication process seems to be on the right track. Here's how it works:

1. The client requests association.
2. The access point sends random cleartext (128-bit challenge).
3. The client encrypts challenge (the nonce) and sends the result.
4. The access point verifies the encrypted challenge.
5. The access point authenticates the client and sends a positive response and then associates the client.

The IEEE 802.11 specification does not require shared-key authentication.

I Know Your Secret

If we asked you which authentication scheme was more secure, we venture you would answer shared-key authentication. Logically, you might think shared-key authentication is more secure than open-system authentication. Oddly enough, you'd be wrong. Because of the way the shared-key authentication works, it's actually *less* secure. The math shows why: An attacker starts to gather management messages from the authentication process. One message contains the random challenge in cleartext. The next message contains the encrypted challenge, using the shared key. Hey, no problem for the hacker; the RC4 algorithm is not complex: The algorithm does an exclusive OR operation on the plaintext to derive the ciphertext, as follows:

```
P XOR R = C
```

Uh-oh. And if you're with us so far, then the rest is just simple math:

```
If P XOR R = C then C XOR R = P
If P XOR R = C then C XOR P = R
```

where P = plaintext, C = ciphertext, and R = key stream (or random bytes). Now the attacker knows everything: algorithm number, sequence number, status code, element ID, length, and challenge text. It's the attacker's turn, and here's how it looks, blow by blow:

1. The attacker requests authentication.

2. The access point responds with a cleartext challenge.

3. The attacker uses the challenge with the value R (as just shown) to compute a valid authentication-response frame by XORing the two values together. Result: He can compute a valid CRC value.

4. The attacker responds with a valid authentication-response message and associates with the AP to join the network.

   The attacker did not need to know the shared-key due to the flaw! (Welcome aboard, stranger. Oops.)

Due to the problems with shared-secret authentication, the standard developers specified WPA and WPA2, both using 802.1X with Extensible Authentication Protocol (EAP).

Have We Got EAP?

So what *is* 802.1X? Did we mean 802.11*x*? No, 802.1X is another IEEE standard, which provides a framework for true user authentication and centralized security management. It provides port level authentication. Initially, the developers offered to standardize security on wired network ports, but others found that the standard had applicability for wireless networking as well.

EAP (Extensible Authentication Protocol) has three components:

✔ **The supplicant:** A client machine trying to access the wireless LAN.

✔ **The authenticator:** A Layer 2 device that provides the physical port to the network (such as an access point or a switch).

✔ **The authentication server:** This verifies user credentials and provides key management.

When the supplicant requests access to an access point, the AP demands a set of credentials. The user then supplies the credentials that the AP may in turn forward to a standard RADIUS (Remote Authentication Dial-In User Service) server for authentication and authorization. RADIUS is commonly used to authenticate dial-in users. However, 802.1X supports the use of an enterprise authentication server or a database service, including a RADIUS server; an LDAP directory; a Windows NT Domain; Active Directory Service or NetWare Directory Service. The exact method of supplying credentials is defined in the 802.1X standard EAP (Extensible Authentication Protocol).

Extended EAP is an addition to the Wi-Fi Protected Access (WPA) and 802.11i (WPA2) certification programs, which further ensures the interoperability of secure Wi-Fi networking products for enterprise and government users. These standards don't specify a set of credentials, so there are several contenders. EAP is, in effect, an authentication "bucket" that allows developers to create their own methods of passing credentials — and it's the main security measure in 802.1X. Following are the six commonly used EAP methods in use today:

- EAP-MD5
- PEAPv0/EAP-MSCHAPv2 or PEAPv1/EAP-GTC (Generic Token Card)
- EAP-Cisco Wireless a.k.a. LEAP
- EAP-FAST
- EAP-TLS
- EAP-TTLS/MSCHAv2

The Wi-Fi Alliance certifies EAP-TLS, EAP-TTLS/MSCHAPv2, PEAPv0/EAP-MSCHAPv2, PEAPv1/EAP-GTC, and EAP-SIM.

Cisco Systems Inc.'s proprietary LEAP (Lightweight EAP) was the first password-based authentication scheme available for WLANs.

So those are the contenders. Regardless of the one you select, you will need an enterprise authentication server. Whether it is ADS, NDS, LDAP-compliant database or RADIUS is up to you.

Let's look at each EAP method.

This method seems easy to digest

The EAP-MD5 method relies on an MD5 hash of a username and password to pass credentials to the authenticator from the supplicant. A message digest (MD) or hash is a mathematical summary of all or any part of a message. In other words, a hash is the transformation of a string of characters into a usually shorter fixed-length value or key that represents the original string.

This method offers no key management or dynamic key generation, thereby requiring the reuse of static WEP keys. (Smell trouble brewing? We do too.)

This method prevents unauthorized users from accessing your wireless network directly, but offers nothing over the proven insecure static WEP encryption scheme. Attackers can still sniff your traffic and decrypt the WEP key. EAP-MD5 also does one-way authentication, there is no mutual authentication — that is, there is no way for the wireless client to verify the access point. Because of this, a determined attacker could plant a rogue access point, called an *evil twin,* on your network and fool your wireless clients into thinking that it is a secure access point. Because EAP-MD5 offers no significant improvements over the standard 802.1X, EAP-MD5 is considered the least secure of all the common EAP standards.

Not another PEAP out of you

Protected EAP (PEAP) is an authentication protocol that uses TLS (Transport Layer Security) to enhance the security of other EAP authentication methods. PEAP for Microsoft 802.1X Authentication Client provides support for EAP-TLS, and uses certificates for mutual authentication, and Microsoft Challenge Handshake Authentication Protocol version 2 (EAP-MS-CHAP v2), which uses certificates for server authentication and password-based credentials for client authentication. Then the client sends the user identity to the authentication server encrypted with the public key of the server. Only the authentication server can read this encrypted information, because it uses its private key.

PEAP works with Active and NetWare Directory Service — that's a bonus — but we're not out of the woods yet. An attacker can trick the authenticator into sending identity or credentials without the protection of a TLS tunnel — allowing the attacker to intercept the information.

Another big LEAP for mankind

EAP-Cisco Wireless or LEAP (Lightweight EAP) was the first password-based authentication scheme available for WLANs. Like EAP-MD5, LEAP accepts a username and password from the wireless device and passes them to the RADIUS server for authentication. What sets LEAP apart from EAP-MD5 are the extra features it adds. When LEAP authenticates the user, one-time WEP keys are dynamically generated for that session. This means that every user on your wireless network is using a different WEP key that no one knows, not even the user. LEAP also stipulates mutual authentication.

LEAP uses MS-CHAPv1 to pass the logon credentials for both the client and access point authentication. MS-CHAP is the challenge handshake authentication protocol developed by Microsoft and used for remote authentication. Unfortunately, the version of MS-CHAP that LEAP uses has known vulnerabilities, and can be compromised by a determined enough attacker with the right tools. You can do off-line dictionary attacks. (Hold on, we'll get there! Look at the use of `asleap` and LEAPcracker later in this chapter.)

That was EAP-FAST

Extensible Authentication Protocol-Flexible Authentication via Secure Tunneling (EAP-FAST) is a publicly accessible IEEE 802.1X EAP type developed by Cisco Systems. Cisco made it available as of February 8, 2004, as an IETF informational draft.

EAP-FAST uses symmetric key algorithms to achieve a tunneled mutual authentication process. The tunnel establishment relies on a Protected Access Credential (PAC) that can be provisioned and managed dynamically by EAP-FAST through an authentication, authorization, and accounting (AAA) server (such as the Cisco Secure Access Control Server).

You can find more information about EAP-FAST at `www.cisco.com/en/US/ products/hw/wireless/ps430/products_qanda_item09186a00802030dc .shtml`.

Beam me up, EAP-TLS

EAP-Transport Level Security (EAP-TLS) is a certificate-based protocol supported natively in Windows XP. Instead of username/password combinations, EAP-TLS uses X.509 certificates to handle authentication. Both the client and the authentication server require certificates to be configured during initial implementation, in other words, mutual authentication. Like LEAP, EAP-TLS offers dynamic one-time WEP key generation, and authenticates the access point to the wireless client as well as the client to the AP. Implementing EAP-TLS requires a public key infrastructure (PKI) to facilitate the handling and sharing of keys. Some organizations may consider this a show-stopper due to its overall cost and complexity.

EAP-TLS provides eavesdropping protection through the use of TLS.

EAP-TLS works with Active Directory using a Microsoft Certificate Server. This method is preferable; especially when running a Win32 client and you are already using certificates. Otherwise you have a steep learning curve to conquer.

EAP-TTLS: That's funky software

You can use EAP-TTLS to provide a password-based authentication mechanism. EAP-TTLS was pioneered by Funk Software as an alternative to EAP-TLS. In EAP-TTLS implementations, only the authentication server is required to have a certificate. Extensible Authentication Protocol-Tunneled Transport Layer Security (EAP-TTLS) authentication uses a two-stage authentication process, which eliminates the need for a certificate on the user side.

The access point still identifies itself to the client with a server certificate, EAP-TTLS establishes the identity of the server using EAP-TLS. The users now send their credentials as username/password. EAP-TTLS then passes the credentials in either PAP, CHAP, MS-CHAPv1, MS-CHAPv2, PAP/Token Card, or EAP form. Sometimes EAP-TTLS is set to use MS-CHAP. Unfortunately, this method is vulnerable to monkey-in-the-middle attacks (which we cover in Chapter 13).

Because EAP-TTLS does not require that you distribute certificates to users, it's a far more convenient protocol to use than EAP-TLS.

The downside to EAP-TTLS is that an attacker can trick the server into sending identity or credentials without the protection of a TLS tunnel.

Implementing 802.1X

To implement 802.1X in your wireless deployment, you have to do some research and be prepared to mix and match vendor's offerings depending on what EAP protocol you plan to use.

First you have to pick a RADIUS server to handle the credential verification. The following list provides different RADIUS servers that can handle some or all of the EAP standards:

- **Microsoft's Internet Authentication Service (IAS):** This is part of Windows 2000, and can authenticate both EAP-TLS and EAP-MD5. If you own a copy of Windows 2000 Server, you already have this RADIUS server, and can access it through Control Panel⇨ Add/Remove Programs. You can find information about IAS at www.microsoft.com/windowsserver2003/technologies/ias/default.mspx.

- **Cisco's Secure Access Control Software (ACS):** This server was developed for both the UNIX and Windows platforms. It is a full-fledged TACACS+ and RADIUS server — and, beginning with version 2.6a, it handles wireless authentication with EAP-Cisco Wireless and EAP-TLS. You can find information about SecureACS at www.cisco.com/en/US/products/sw/secursw/ps2086/.

✔ **Steel Belted RADIUS:** Funk Software is generally known for this product. They have a stripped-down version of that platform (known as Odyssey) that handles wireless authentication exclusively. Odyssey handles the widest range of EAP types by supporting all four of the commonly used protocols, EAP-MD5, EAP-TLS, EAP-Cisco Wireless, and EAP-TTLS. You can find information about Steel Belted RADIUS and Odyssey at www.funk.com/.

✔ **AEGIS:** This product, from Meetinghouse Data, offers a RADIUS server that runs on the Linux platform and handles EAP-TTLS and EAP-TLS authentication. You can find information about AEGIS at www.mtghouse.com/products/aegisserver/index.shtml.

✔ **freeRADIUS:** This is an open-source project that runs on the Linux platform. freeRADIUS supports both EAP-MD5 and EAP-TLS. You can find freeRADIUS at www.freeradius.org/.

The second piece needed to implement 802.1X is an access point that can pass the authentication messages. Because the AP acts only as a conduit for the authentication messages, there are no compatibility clashes between certain access points and different EAP protocols. As long as the EAP protocol fits the standard, then an 802.1X-enabled AP can use it. Generally all enterprise-level APs are either currently capable of handling 802.1X requests either out of the box or through a firmware upgrade for existing equipment. If you have a much smaller environment (or just don't want to spend the money on enterprise class APs), there are hacks available for home-user-level APs to support 802.1X authentication.

The final piece of the 802.1X puzzle is the client software. Again, depending on your OS and preferred EAP standard there are several to choose from, such as:

✔ **Cisco:** Cisco writes its ACU client piece to make its adapters work with EAP-Cisco Wireless on all flavors of Windows, Apple, and Linux.

✔ **Windows XP:** Windows XP has a built-in client for EAP-TLS and EAP-MD5. It only works properly with Microsoft issued certificates though. This client is supposed to be released for Windows 2000 and CE.NET as well.

✔ **Odyssey:** Funk's Odyssey client currently works on all flavors of Windows except CE. Linux, CE and Apple versions are all due out in the next few of months. The Odyssey client handles EAP-MD5, EAP-TLS and EAP-TTLS, and will soon offer EAP-Cisco Wireless support for non-Cisco hardware.

✔ **AEGIS:** The AEGIS client from Meetinghouse Data offers EAP-TLS, EAP-TTLS, and EAP-MD5 support for all flavors of Windows (except CE) and Linux.

✔ **Others:** Several open-source clients are being worked on currently, most notably Xsupplicant from the open1X project.

To use 802.1X, do the following:

- ✔ Set the authentication method to Open.
- ✔ Have your broadcast keys rotate every ten minutes or less.
- ✔ Use 802.1X for key management and authentication.
- ✔ Look over the available EAP protocols and decide which is right for your environment.
- ✔ Set the session to time out every ten minutes or less.

You can find a list of open EAP issues at www.drizzle.com/~aboba/EAP/eapissues.html. Also, you can find a commercial EAP testing tool from QA Cafe (www.qacafe.com/suites-eapol.htm). You can use its EAPOL to test your 802.1X authentication. EAPOL is a serious lab-testing tool and is not really intended for "script kiddies." It provides test coverage of the EAPOL protocol along with specific EAP-MD5, EAP-PEAP, EAP-TLS, and EAP-TTLS tests. Check out the site and you'll find a demo version for Linux.

Let's look at a major EAP problem: cracking LEAP.

Cracking LEAP

Implementations of 802.1X with LEAP established a strong foothold in the enterprise market. Because LEAP was one of the first solutions available, crackers wrote LEAP crackers. As such, LEAP represents a large security vulnerability for most enterprise wireless LANs. Even so, few enterprises seem to care.

The LEAP weakness was well known from the beginning, because LEAP is essentially an enhanced version of EAP-MD5 with Dynamic Key Rotation and Mutual Authentication. Part of the problem is that LEAP relies on MS-CHAPv2 (Microsoft Challenge Handshake Authentication Protocol version 2) to protect the authentication of user credentials. MS-CHAPv2 is weak because it does not use a salt for the NT hashes, uses a weak 2-byte DES key, and sends usernames in cleartext. So LEAP inherits the following MS-CHAP flaws:

- ✔ Cleartext username
- ✔ Weak challenge/response DES key selection: a 8-bit challenge, hashed with MD4 (Can you say, "You've got to be kidding?")
- ✔ Absence of a salt for the stored NT hashes

Because LEAP is susceptible to off-line dictionary and brute-force attacks, you should not use LEAP for secure networks. Cracking LEPA is made easier by maintaining a 4-terabyte database of likely passwords with pre-calculated hashes. What most users think is a strong password is usually really weak and breakable within minutes. An attacker can do this with relative impunity and zero chance of detection, since the attack is passive and performed off-line. Cisco feels that you can make LEAP secure by increasing the complexity of the password, thereby thwarting off-line dictionary and brute-force attacks. Although this is true, the possibility that someone actually *will* create a ten-character, uppercase-and-lowercase, alphanumeric password peppered with special characters is (to put it mildly) slight. Think about the passwords in your organization. Let's face facts: Any password you expect people to remember is easily cracked using a dictionary or brute-force attack.

For an attacker who's in a hurry (and most of them are), cracking LEAP is far more productive than cracking the infamously weak WEP protocol. You can usually crack LEAP in several minutes, compared to the hours it might take to crack WEP. So LEAP is a definite target.

There are several LEAP solutions available, including these gems:

✔ `asleap`

✔ THC-LEAPcracker

✔ `anwrap`

These are discussed in detail in the following sections.

Using asleap

Should you want to test your implementation of LEAP to see whether your organization uses strong passwords, you can use `asleap` from Joshua Wright. This tool makes it easy to capture the required login traffic by allowing you to spot WLANs that are using LEAP — and then de-authenticate users on the WLAN, forcing them to reconnect and reenter their usernames and passwords. You'll find that weak passwords fall rapidly when pitted against a tool such as `asleap`.

`asleap` allows you to scan the wireless-network broadcast spectrum for networks that use LEAP, capture wireless network traffic, and crack user passwords. `asleap` is a busy little program. Here's a quick look at what it does:

✔ Recovers weak LEAP passwords

✔ Reads frames from any wireless interface running in RFMON mode

✔ Monitors a single channel, or hops channels to look for target networks that are using LEAP

✔ Actively de-authenticates users on LEAP networks, forcing them to re-authenticate, which makes the capture of LEAP passwords very fast

✔ Only de-authenticates new users, doesn't waste time on user accounts that aren't running LEAP

✔ Reads from stored libpcap files or AiroPeek NX files

✔ Reads live from any Ethernet network interface

✔ Uses a dynamic database table and index to do lookups on large files very rapidly

✔ Cracks PPTP VPN authentication sessions that use MS-CHAP

Figure 15-1 shows the syntax for `asleap`.

Figure 15-1:
The asleap
syntax.

Should you want to find more information about asleap, check out the mailing list at `http://lists.sourceforge.net/lists/listinfo/asleap-users`.

The source and Win32 binary distribution are available at `http://asleap.sourceforge.net`. The latest version does PPTP as well as LEAP captures. `asleap` is released under the GNU Public License (GPL).

Using THC-LEAPcracker

The THC-LEAPcracker Tool suite contains tools to break the NTChallenge-Response encryption technique used by Cisco Wireless LEAP Authentication. Also included are tools for spoofing challenge packets from Access Points, so you can perform dictionary attacks against all users.

You can find THC-LEAPcracker at `http://thc.org/releases.php?s=4&q=&o=`.

Using anwrap

Written by Brian Barto and Ron Sweeney, `anwrap` is a wrapper for `ancontrol` that serves as a dictionary-attack tool against LEAP enabled Cisco Wireless Networks. It traverses a user list and password list, attempting authentication and logging the results to a file. `anwrap` causes havoc on NT Networks that have lockout policies in place.

`anwrap` requires ancontrol and Perl. The `ancontrol` command controls the operation of Aironet wireless networking devices via the an driver. The `anwrap` author tested the tool on FreeBSD 4.7.

You can find `anwrap` at `http://packetstormsecurity.nl/cisco/anwrap.pl`.

As a result of cracker tools like `asleap`, THC-LEAPcracker and `anwrap`, Cisco has de-emphasized the use of LEAP, especially for those organizations that can't or won't enforce strong passwords. They now recommend the use of EAP-FAST.

Network Authentication Countermeasures

If you had your heart set on a life of carefree wireless-network use, maybe you're ready to put your head in the oven and turn on the gas. Don't do it. There are some things you can do to protect yourself. Help is on the way.

WPA improves the 8021.1 picture

Because of the WEP problems, the IEEE approved Wi-Fi Protected Access (WPA) as an interim solution to address those problems. WPA is an example of a software or firmware patch and does not require the hardware upgrade that 802.11i does.

The objective of WPA was to bring a standards-based security solution to the marketplace to replace WEP until the availability of the full-blown IEEE 802.11i Robust Security Network (RSN), an amendment to the existing wireless LAN standard.

Two key features WPA are its most significant improvements:

- ✓ **802.1X support:** WPA uses 802.1X port access control to distribute per-session keys. Some vendors previously offered 802.11X support even though it was not specified in the standard. The 802.1X port-based access control provides a framework to allow the use of robust upper-layer authentication protocols.

- ✓ **Temporal Key Integrity Protocol (TKIP):** WPA uses the Temporal Key Integrity Protocol (TKIP) to address WEP problems such as IV length and key management.

But WPA is not without its problems. Basically, one can crack Wi-Fi Protected Access Pre-Shared Keys that use short dictionary-word–based passphrases. You will find software to help with this as well. The WPA Cracker (`www.tiny peap.com/page8.html`) tool is somewhat primitive, requiring that you enter the appropriate data retrieved via a packet sniffer. The author recommends you use ethereal.

Joshua Wright, who wrote `asleap`, offers us CoWPAtty (`http://new.remote-exploit.org/`), which is another off-line WPA-PSK–auditing tool.

For WPA, certain shorter or dictionary-based keys are easy to crack because an attacker can monitor a short transaction or force that transaction to occur and then perform the crack remotely.

So what do you do? Well, you can:

- ✓ Choose better passphrases, especially ones that aren't made up of words in the dictionary. Select passphrases that are random and at least 20 characters in length.
- ✓ Use WPA Enterprise or 802.1X with WPA.

Alternatively, you can use virtual private network technology, such as those technologies described below.

Using WPA2

WPA is still based on the RC4 algorithm, a stream cipher. But a major component of new RSN specification in 802.11i is the use of the Advanced Encryption Standard (AES) for both data confidentiality and integrity.

We strongly recommend you look for and implement technology that supports 802.11i. The 802.11i specification offers Advanced Encryption Standard (AES)-based data link level cryptographic services that are validated under FIPS 140-2. The ratified standard WPA2 uses the AES-CCMP (Counter Mode-Cipher Block Chaining MAC Protocol) algorithm.

The AES-based solution will provide a highly robust solution for the future but will require new hardware and protocol changes. (For more about the advantages of 80211i, WPA and AES, see Chapter 14.)

Using a VPN

Your organization may find that it is necessary to use a virtual private network (VPN) scheme. A VPN helps against the risk of eavesdropping by providing an encrypted tunnel between two networks that only authorized persons can access. A tunnel is created using an accepted technique between two endpoints. Any data traveling between those two points is secured using encryption. The tunnel is set up and torn down each time you use it. (For more about VPNs, see Chapter 14.)

The solution you decide upon needs to fulfill your business needs. There are many different implementations of virtual private networking. These range from commercial third-party applications to those that are embedded in operating systems. There are several potential VPN solutions as follows:

- Point-to-Point Tunneling Protocol (PPTP)
- Layer 2 Tunneling Protocol (L2TP)
- IPSec
- SSH2

You should know that people were kind enough to release other PPTP crackers (in addition to `asleap`), such as these:

- **Anger:** Anger, which is a PPTP MS-CHAP challenge/response sniffer that can feed output to `L0phtcrack`. You can find Anger at `www.securiteam.com/tools/6F00X000AU.html`.

- **Deceit:** Aleph One released deceit, which you can find at `http://packetstormsecurity.nl/new-exploits/deceit.c`.

- **Ettercap:** Don't forget to look at ettercap (`http://ettercap.sourceforge.net/`). Ettercap has plug-ins to sniff PPTP tunnels, decapsulate traffic, and retrieve user passwords. If that makes you think you should use something other than PPTP, you got it in one!

But just when you think another protocol is secure, you find out it isn't. There are CheckPoint VPN-1, Cisco VPN Client, Nortel Contivity VPN Client, OpenBSD isakmpd, PGPFreeware, SafeNet, and WAVEsec versions of IPSec that are susceptible to monkey-in-the-middle and buffer overflow attacks. You can use `ike-scan`, `IKEProbe`, `ipsectrace`, and `IKEcrack` to test those

IPSec VPNs. The first three are available from `www.forinsect.de/pen test/pentest-tools.html`. You can find `IKEcrack` at `http://ikecrack. sourceforge.net/`. You should also try `kracker_jack` to perform a monkey-in-the-middle attack. `kracker_jack` is part of AirJack (`http:// sourceforge.net/projects/airjack/`).

WIDS

Wireless intrusion detection (as detailed in Chapter 11) requires its own system. AirSnare is another tool to add to your Wireless Intrusion Detection System toolbox. AirSnare will alert you to unfriendly MAC addresses on your network — and will also alert you to DHCP requests taking place. If AirSnare detects an unfriendly MAC address, you can track that MAC address's access to IP addresses and ports — or launch `ethereal`.

Figure 15-2 shows AirSnare running. You can see that it identified some "unfriendly MAC addresses."

You can find AirSnare at `www.majorgeeks.com/download4091.html`.

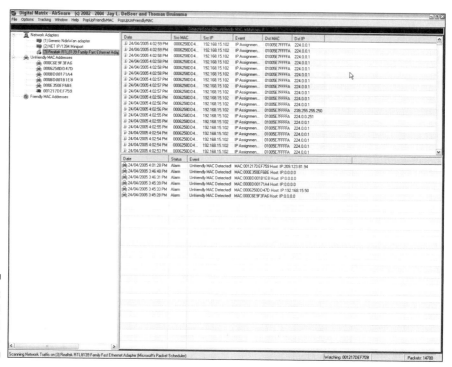

Figure 15-2:
AirSnare
finding
intruders.

Use the right EAP

As you saw there are several versions of EAP. Which one do you use? You should select one that is non-proprietary. Also, make sure your solution provides protection from a variety of network attacks, including man-in-the-middle, authentication forging, weak-IV attacks (AirSnort), packet forgery (replay attacks), and dictionary attacks. Obviously, your solution should support a variety of user- and password-database types, support password expiration and change, and be flexible, easy to deploy, and easy to manage.

Setting up a WDMZ

You must treat any wireless segment as untrusted — and because it's an untrusted segment, you must protect yourself. Generally we accomplish this through the use of a security method called a *wireless demilitarized zone* or *WDMZ*. The term DMZ has its roots in geopolitics: It is the "no man's land" between North and South Korea. Take a stroll in the DMZ, and both sides will take a shot at you. With WDMZs, we won't shoot you, but we won't trust you.

You should add a wireless DMZ or screened wireless network between the internal network (that is, your intranet) and the external network (the Internet). This DMZ is where you put your access point, authentication server, Web server, and external DNS server. You use an authentication server to regulate traffic between the untrusted network (the Internet) and the trusted network (the intranet). With wireless, we compartmentalize our networks or segregate our access points — and have a trusted way into the internal network. Bottom line: You should segregate wireless stations onto *one or more separate network segments* and prevent any direct, unauthenticated communication with other devices on the wired portion of the network. By compartmentalizing, we can isolate risks and apply controls to mitigate or eliminate the risk.

In summary, you must not allow anyone to place an access point behind your firewall. Place the access point on a segment that the firewall filters.

Using the Auditor Collection

In this book, we showed you a lot of tools. In Chapter 8, we mentioned the Auditor Collection. You can find the Auditor Collection at `http://new.remote-exploit.org/index.php/Auditor_main`. The Auditor Collection is an ISO image of Knoppix that includes over 300 tools. It is a "must-have" for anyone doing ethical hacking in an organization — especially the wireless variety. It provides the most useful Linux, wireless, ethical-hacking tools. Fortunately it's really easy to use.

Auditor installs itself and a complete Linux 2.4.9 kernel on a RAM disk — and executes in RAM. Auditor has complete PC Card support and is ideally run on a laptop computer. You can mount a hard drive or floppy drive and save logs and reports.

Using the Auditor CD-ROM is as easy as the following steps.

1. **Open your laptop CD-ROM bay. Then power down Windows or Linux or whatever operating system you are using.**

2. **Insert the Auditor CD-ROM into the drive and close it.**

3. **Power up the computer and watch Auditor boot.**

   If this is the first time you have used Auditor, interrupt the boot process and enter the BIOS set-up program. Make sure that your system will boot from the CD-ROM before it tries the hard drive. Obviously the exact method for doing this depends on your hardware manufacturer and the BIOS you're using, but follow the on-screen instructions to enter the setup program. When you're done, you won't need to do this again.

   You see an Auditor splash screen with license and credit information.

4. **Should Auditor pause temporarily at a** `boot:` **prompt, select the appropriate screen resolution; then hit Enter or Return.**

   Eventually you see the Auditor desktop, as shown in Figure 15-3: a picture of a centurion.

   In the lower-left portion of the window, you see a K with a gear. This is the KDE manager icon.

5. **Click the KDE manager icon.**

   A menu pops up.

6. **From the menu, select Auditor⪢Wireless⪢LEAP/PPTP cracker⪢ ASLeap (LEAP/PPTP cracker), as shown in Figure 15-4.**

   This is Auditor's ASleap tool. The window shown in Figure 15-1 opens.

Take some time and explore all the programs available with the Auditor CD-ROM. Whenever you want to use Auditor, just put the disk in the drive and turn on the power. When you're finished with Auditor, shut down, remove the Auditor CD-ROM, and reboot. Your system will boot whatever operating system it gets from your hard drive (assuming you have set it up that way).

Now you have a handful of wireless networking ethical-hacking tools. So you have no excuse. Dedicate a laptop to ethical hacking and get cracking.

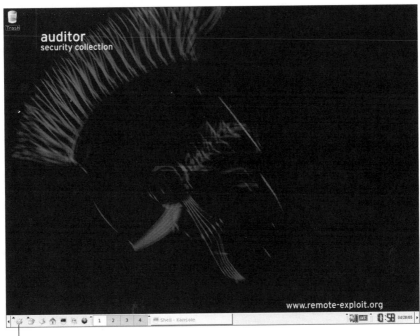

Figure 15-3:
Auditor
desktop.

KDE Manager icon

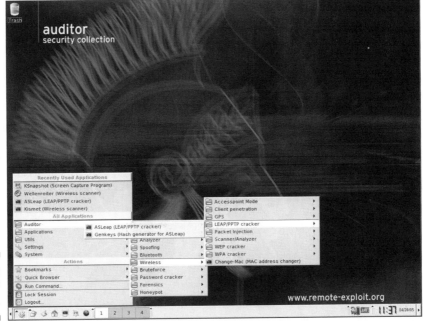

Figure 15-4:
LEAP
crackers.

Part IV
The Part of Tens

The 5th Wave   By Rich Tennant

"Ironically, he went out there looking for a 'hot spot'."

In this part . . .

We made it — and we're so glad you've joined us! We're at the end of the ethical wireless-hacking road — if there truly is a such a thing. In this part, we put together some great wireless hacking resources that you can benefit from in the road ahead. We not only recap our favorite wireless-hacking tools, but we also talk about common wireless-hacking mistakes you can avoid. We also outline critical steps you need to perform after you've finished your wireless-security tests. Keep this part handy — this is some of the most important material of the entire book.

Chapter 16

Ten Essential Tools for Hacking Wireless Networks

As with any trade, it's essential to have the right tools when testing your wireless network for security vulnerabilities. Here are ten tools we have found that get the job done.

Laptop Computer

For starters, you've got to a have a good test system — preferably a portable laptop computer. Although it is possible to perform wireless-security testing using a handheld device such as a Pocket PC, the tools available on such devices are limited compared to those on a laptop system.

Due to the multiple operating system requirements of the popular wireless testing tools, we recommend using either a system that can dual boot Windows (preferably 2000 or XP) and Linux (any recent distribution will do) or a Windows-based system running a virtual machine program (such as VMware) on which you can install multiple operating systems. The hardware requirements for systems running a single operating system are pretty minimal given today's standards. A system with a Pentium III or equivalent processor, 256MB RAM, and at least a 30–40GB hard drive should be more than enough. If you'll be running VMware or another virtual machine program, you'll want to at least double this amount of RAM and hard drive space.

Wireless Network Card

In addition to the laptop, you've got to have a good wireless network-interface card (NIC). Look for a PC Card NIC that's not only compatible with the various wireless tools, but one that also has a connector for an external antenna so you can pick up more signals. The Orinoco Gold card (and its re-badged equivalents) serves both purposes very well. Many wireless NICs built in to today's laptops are good general purpose cards, but your test results may be limited due to the shorter radio range capability of the internal antennas.

Antennas and Connecting Cables

A high-gain unidirectional or omnidirectional antenna — or cantenna — will do wonders for you when you're scanning your airwaves for wireless systems. When you're shopping for antennas, look for one with a pigtail connection that matches the type of connector you have on your wireless NIC. Also be aware that the length of these pigtail cables should be kept as short as possible. Because they're made with a very thin microwave coax, these cables have fairly high signal losses at microwave frequencies and with the connectors placed on either end of the pigtail cable. To avoid high cable losses, you should not use a pigtail cable longer than 5 feet.

GPS Receiver

If you'll be war-walking/driving/flying — or if your wireless systems span across a large building or campus environment — then it's time to think globally: A global positioning satellite (GPS) receiver will come in handy. With a GPS receiver, you'll be able to integrate your wireless testing software and pinpoint the locations of wireless systems within a few meters.

Stumbling Software

To get your wireless testing rolling, wireless stumbling software is essential; you can use it to map out things like SSIDs, signal strength, and systems using WEP encryption. Software you can use for this includes Network Stumbler for Windows or your wireless NIC management software. For really basic stumbling, you can even use the management software built in to Windows XP.

Wireless Network Analyzer

To probe deep into the airwaves, a network analyzer is essential. Programs such as Kismet, AiroPeek, and `ethereal` can help you monitor multiple wireless channels, view protocols in use, look for wireless system anomalies — and even capture wireless data right out of thin air.

Port Scanner

A port scanner such as `nmap` or SuperScan is a great tool for scanning the wireless systems you stumble across to find out more about what's running and what's potentially vulnerable.

Vulnerability Assessment Tool

A vulnerability-assessment tool such as Nessus, LANguard Network Security Scanner, or QualysGuard is great for probing your wireless systems further to find out which vulnerabilities actually exist. This information can then be used to poke around further and see what the bad guys can see and even potentially exploit.

Google

It's not only a great reference tool, but the Google search engine can also be used for searching Network Stumbler .NS1 files, digging in to the Web-server software built in to your APs, finding new wireless-security testing tools, researching vulnerabilities, and more. The Google taskbar (downloadable for Internet Explorer, built in to FireFox) makes your searching even easier.

An 802.11 Reference Guide

While performing ongoing ethical hacks against your wireless systems, you'll undoubtedly need a good reference guide on the IEEE 802.11 standards at some time or another. The 802.11 wireless protocol is very complex and will evolve over time. You'll likely need to look up information on channel frequency ranges, what a certain type of packet is used for, or perhaps a default 802.11 setting or two. The Cheat Sheet, the wireless resources found in Appendix A in this book, as well as Peter's book *Wireless Networks For Dummies* are good references that can really help you.

Chapter 17

Ten Wireless Security-Testing Mistakes

From our experience, ethical hackers tend to make the same mistakes over and over again. Perhaps they learned from each other. We see no reason for you to make these same mistakes. In no particular order, the top ten security testing mistakes are as follows:

Skipping the Planning Process

As we stress in Chapter 2, planning is a tedious but extremely important task. You must clearly define the boundaries of your work to avoid any conflict or question of misdeed. Despite all attempts at thoroughness and efficiency in your testing, one of the largest factors in determining the success of a security test is the thoroughness of your planning. The planning process is handled far from the tester's toolbox. It requires a certain level of project management and communication skills. Your plan is the company's official declaration of what it wants to accomplish and how it wants to do it. Remember: Very few people ever arrive at their destinations without first intending to get there.

At a minimum, your plan should specify the following:

- ✔ The roles and responsibilities for everyone involved in the ethical hack.
- ✔ The level of involvement of each tester and the importance of her participation in the team.
- ✔ The schedule for when the testing will take place. Management may prefer that the testing be done when traffic is low, which might translate into late nights, early mornings, or weekends.

Your security defense budget is likely small, so you need to operate with efficiency and creativity to do more with less. So plan carefully and meet the expectation of the plan. Otherwise, in the future, your management or customer may see security testing as an unnecessary cost.

Not Involving Others in Testing

Often the trick to a successful test lies in observing the details. One such detail is the inclusion of other individuals from your organization in the testing process. Talk to your network professionals. Get people involved up-front during planning. They may help you save time or money by providing insight into the network that you might not have.

Ensure that the testing process is closely monitored by others. Involving others during the process may save you reporting time or may save your hide if you're accused of something you did not do.

Not Using a Methodology

Ethical hacking is different from penetration testing. Ethical hacking is extremely methodical and relies on a method. In Chapter 2, we discuss the concept of the scientific method. You need to adopt or develop a method. Your method should consist of the following steps: planning, testing, and reporting. The method may consist of best practices, such as Open-Source Security Testing Methodolgy Manual (OSSTMM) and Information System Security Assessment Framework (ISSAF). (For more on these terms, refer to Chapter 2.)

Although having a testing methodology is important, don't constrict the creativity of the tester by introducing strict methods that affect the quality of the test. Be flexible and leave some tasks open to interpretation.

Forgetting to Unbind the NIC When Wardriving

Later on in this chapter, we talk about the legalities of ethical hacking. For now, we want to avoid doing anything that someone might perceive as illegal or unethical. At no time should you connect to any access point that isn't yours. One of the things you can and should do is ensure that your wireless adapter does not authenticate with an access point you stumble upon. This is called *autoconnection* or *accidental association*. Avoiding autoconnection is essential.

Some client adapters are more likely than others to connect with any open access point that comes into range, given enough time to perform a DHCP transaction.

To avoid autoconnection, sometimes you must simply disable your client adapter's client utility before setting out on a wardrive. But that may not work. The only foolproof way to prevent autoconnection is to disable all networking protocols, that is, TCP/IP, NetBEUI, NetWare, and other communication protocols on your wardriving computer. Without a networking protocol, the computer cannot communicate. However, even without a networking protocol, the Wi-Fi client adapter still gathers frames and NetStumbler, or Ethereal can display them.

Your local law enforcement organization may interpret the simple act of associating as computer trespass, so you need to ensure you don't accidentally associate. The solution is simple. All you need do is unbind the TCP/IP protocol from the wireless adapter.

With accidental association, not only do you have a potential legal problem, but you also have a potential wardriving problem as well. Client adapters that autoconnect to an access point sometimes place the SSID of that access point in the SSID field of the wireless adapter's operating profile. From that point forward, your wardriving program will not log any additional stations with any other SSIDs. This effectively ends your wardrive, even when you drive around different neighborhoods for a long time.

If you are a DHCP client, then you can use the Windows `ipconfig.exe` command to release all current TCP/IP network configurations. Follow these steps:

1. **Select Start⇨Run.**

2. **Enter** `command` **in the Open dialog box in the Run window.**

3. **At the command prompt, type** `ipconfig /release_all`.

   This releases all IP addresses and sets them to `0.0.0.0`. If you want more information about this command, type `ipconfig /?`.

If you still use Windows 98/98SE or Windows ME, you can use the built-in `WINIPCFG.exe`. It is easy to use this program, but if you're not sure, read the help file. If you're a Linux user, you can use the `ipconfig /renew` command.

You can get a graphical version of `ipconfig` by downloading `wntipcfg.exe` from `www.microsoft.com/windows2000/techinfo/reskit/tools/existing/wntipcfg-o.asp`. Figure 17-1 shows `wntipcfg` in action. It's simple to use: To release all IP addresses and set them to `0.0.0.0`, just click the Release All button. For more information about how to use `wntipcfg`, Microsoft has excellent help on the download page for the care and feeding of the program.

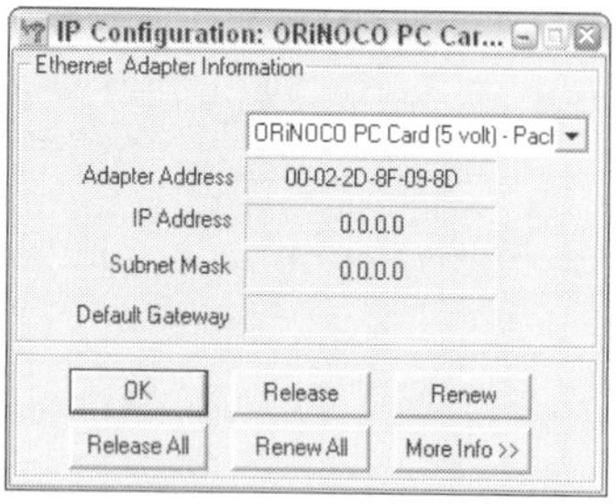

Figure 17-1:
Using
`wntipcfg`
to release IP
addresses.

> **IP Configuration: ORiNOCO PC Car...**
> ┌ Ethernet Adapter Information ──────────────
> │   ORiNOCO PC Card (5 volt) - Pacl ▼
> │   Adapter Address   00-02-2D-8F-09-8D
> │   IP Address   0.0.0.0
> │   Subnet Mask   0.0.0.0
> │   Default Gateway
> │
> │   [OK]   [Release]   [Renew]
> │   [Release All] [Renew All] [More Info >>]

If you are going to use MiniStumbler or CENiffer to release the IP addresses, you need to manually change the IP address to an unusable address. To change the IP address in Pocket PC, follow these steps:

1. **Choose Start⇨Setting⇨Network Adapters.**

2. **Select your wireless card and click Properties.**

3. **Select the Use Specific IP Address radio button and enter** `0.0.0.0` **for the IP address, Subnet mask, and Default gateway.**

The only way to prevent autoconnect in Windows XP is to disable the TCP/IP protocol from the wireless card or adapter before you go wardriving. Without the TCP/IP protocol, your client can connect but not communicate. How you disable networking protocols depends on your operating system. For Windows XP, it's easy to do:

1. **Choose Start⇨Control Program.**

2. **Double-click Network Connections.**

   You should see a window similar to the one in Figure 17-2.

3. **Highlight the wireless network connection.**

4. **From the Network Tasks list, select Change Settings of This Connection.**

   You should see a window like the one shown in Figure 17-3.

5. **Scroll down in the list of installed protocols until you see Internet Protocol (TCP/IP). Highlight this item.**

6. **To the left, you will find a check box. Uncheck the box.**

7. **Click OK.**

8. **Restart your computer.**

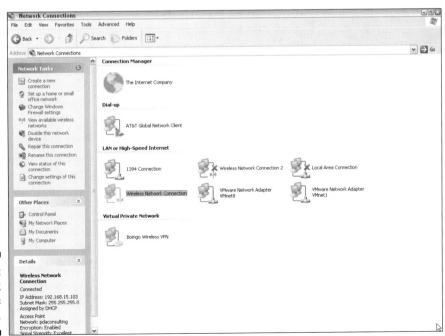

Figure 17-2:
Network
Connections
window.

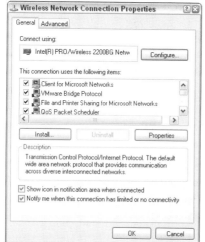

Figure 17-3:
Wireless
Network
Connection
Properties
window.

When you restart your computer, you can still use your wireless card; that is, the card is not disabled, but the TCP/IP protocol is no longer bound to the card. You cannot use the protocol to send packets.

Enabling and disabling networking protocols is a lot of work, so we recommend getting a dedicated machine for your ethical hacking.

Failing to Get Written Permission to Test

Performing security tests of any kind against any network without explicit written permission from the appropriate authority is not a good thing. Here are some reasons why:

- ✔ When you undertake an ethical hack, you may perform some actions that are similar, if not identical, to those carried out by a real attacker. Without documentation, how can anyone tell the ethical hackers from the not-so-ethical ones?

- ✔ Your testing may result in the compromise of the network.

- ✔ You shouldn't do Denial-of-Service tests without explicit permission. Most organizations frown on you bringing down the system and preventing people from doing their work.

- ✔ Even when you have an agreement, your client or boss may not accept, for instance, that you may uncover classified information or gain access to internal resources.

For these reasons, you should always obtain clear authorization from the appropriate authority to perform your ethical hack. Notice the use of the word *appropriate* in the prior sentence. It is very important that the right person has given permission to you. You should obtain authorization from someone who has the authority to approve your work.

Regardless of whether you work for the company or you are an external consultant, make sure the approval is printed on company letterhead and signed by the responsible person. If you are an external consultant, make sure you have a signed contract.

 One last thought on Denial-of-Service testing. As a rule, security assessment methodologies such as OSSTMM or ISSAF do not require the tester to try a Denial-of-Service exploit. The tester should only use gathered evidence to provide a proper review of security processes and systems. So, unless you have a compelling reason to do so, you should not perform Denial-of-Service tests.

Failing to Equip Yourself with the Proper Tools

Just like any other professional or tradesperson, you rely heavily on your tools. You must make sure your tools are kept tuned and ready to go. You should know where the tools came from, how the tools work, and test them in a restricted test area before using the tools for an ethical hack. You should not use a new or unfamiliar tool on a production system or network. Don't trust your providence to unsigned tools. If the developer signed the software with a GPG (www.gnupg.org) or other signature, you should verify the signature by using the correct key. Otherwise, you cannot be sure the program is not a Trojan horse itself.

Authors frequently update their software to fix security and other problems. So ensure you're using the latest version of your tools. For example, you can download the Kismet development code by typing the following at the command prompt:

```
svn co http://www.kismetwireless.net/code/trunk kismet-devel
```

This command fetches the latest development version for you.

To protect yourself, your information, and your network, you should use dedicated equipment for your testing. Sending e-mail and performing other administrative or personal tasks on your testing machine is just asking for trouble. You don't want a virus or Trojan horse on a system you use for security testing. Do these tasks on another machine. When you have access to absolutely only one machine, use different boot partitions. (Consider using Symantec's PartitionMagic, which we cover in Chapter 4.) This guarantees the integrity of your testing machine.

Another action you can take to protect the integrity of the machine is to make sure it is fully patched. Install the latest patches for the operating system and any applications installed on the tester's machine. To keep the cruel and unforgiving world at bay, implement a host-based firewall and anti-virus software.

Ensure that your team has licenses for all the commercial tools you use. If you decide to use freeware or open-source tools, make sure the agreement or license allows you to do so. Some of the developers of free software don't want you using their software for commercial gain. So make sure all your tools are legal, updated, and fully licensed.

Over-Penetrating Live Networks

Ethical hacking follows no steadfast rules, but there are some commonsense rules to consider. One is don't test until you negatively impact networks or systems. Hopefully, you are testing at times that won't impact clients of the network, so perhaps this is not an issue. Always stay professional and you should not have any ethical problems such as negatively impacting your target.

Using Data Improperly

You can share the data from your ethical hack with like-minded people and your boss or customer, you can show it to anyone who can change things for the better, and you can use it for education, but don't use the data for personal gain.

Don't use anything you learn to embarrass anyone — especially your boss.

Failing to Report Results or Follow Up

The reporting phase is as important as the testing itself. Professionals know that reporting is critical and takes time. Remember to include enough time in your budget for the reporting process. Writing a great report requires a great deal of effort. Peter has found that writing and getting agreement to a formal report, takes three times as long as the work itself. It really doesn't matter how good your work is if you cannot tell the story well.

Also, report the risk vulnerabilities you've discovered *promptly*. Don't wait until someone exploits the vulnerability or until you report your findings. Your company or customer would have a difficult time proving due diligence if you knew about the vulnerability but did not report it. In these circumstances, not only should you report these items to your boss or customer, but you must also present them with a practical solution.

Before you start writing your report, plan the activities you need to prepare and submit the report. Plan to share your findings with all those with an interest, such as network administrators and your sponsor. You should also plan to share a draft with people. These steps take time.

Your report is one way for you to show the completeness and rigor of your testing methodology. Your peers can review your methods, your findings, your analysis, and your conclusions and decide whether you came to the correct conclusions based on what you report. Some thoughts on reporting:

- ✔ Reports should include the following sections:
  - Executive Summary
  - In Scope Statement
  - Out of Scope Statement
  - Objectives
  - Nature of the Testing
  - Analysis
  - Summary of Findings
  - Vulnerability Summary
  - Countermeasure(s) to Control the Vulnerability
  - Conclusion
  - Supporting Documentation
- ✔ Reports should include all assumptions regarding the network or system under review.
- ✔ Reports should include all unknowns, and they should be clearly marked as unknowns.
- ✔ Reports should state clearly all states of security found, not only failed security measures.

✔ Reports should be based on the established deliverable.

✔ Reports must use only qualitative metrics for gauging risks based on industry-accepted methods.

✔ Reports should include practical solutions toward discovered flaws.

Many organizations make the major mistake of not following up swiftly. It's one thing to identify vulnerabilities; it's another thing altogether to fix the problem.

If you don't intend to fix the problems promptly, then don't bother with the assessment. You are definitely in a worse legal position if you don't fix known problems than if you don't know about the problems. In this case, ignorance is bliss — or judicious.

Failing to follow up on the findings of the testers has a negative effect on the testers as well: It's demoralizing. Nothing is worse than doing your job and doing it well and discovering that nobody notices or cares. Because you have placed a lot of trust in your ethical hackers, not valuing their work might not be such a good tactic.

Finally, if you don't follow up without delay, you also risk wasting the money you spent on the test itself. Most information technology environments are extremely dynamic; you may have to repeat the tests if you don't follow up straightaway.

Breaking the Law

You get in your car and drive around the neighborhood. But you're not running errands. Instead, you're using your laptop with Windows XP, NetStumbler, an omnidirectional antenna, an ORiNOCO Wi-Fi card, and your Pharos GPS device to collect status information and the location of wireless networks. As you drive, NetStumbler records open networks it finds. After you finish driving around the neighborhood, you map the location of open and closed networks and upload the map to the Internet. Given this scenario, the following questions arise:

✔ **Is this illegal?** Conceivably, this is trespass and illegal access.

✔ **Is this unethical or unprofessional?** Technology is not bound to ethics. It is the application and use or abuse of technology that brings the question of ethics in.

✔ **Is wardriving harmless fun or a computer crime?** Like hacking itself, the answers are not black and white. There are shades of gray or inconspicuous ethical shades to wardriving that most people don't really understand. Wardriving in and of itself is quite innocuous, legal, and benign. It is what you do with the information that is important.

In 2004, news programs on many U.S. television stations carried stories about wardriving. In these stories, a local security "expert" shows the reporter how easy it is to connect to a wireless access point. In one particular segment, a Minneapolis-St. Paul TV station showed the "expert" actually sending something to print on the compromised network. One would think that at best that such activity is unethical, if not illegal. However, the legality of wardriving is untested. Regardless of whether wardriving is legal, as a professional, you should avoid any perception that you are taking part in illegal or unethical behavior. Remember that if you are doing the ethical hacking on behalf of your company, then your behavior reflects on the company.

Grove Enterprises Inc. (`www.grove-ent.com/LLawbook.html`) has published some information on the legal aspects of wardriving on its Web site. There are some really interesting and intriguing facts on the site. Most Americans are ill-informed of the pertinent laws. The laws in some states are so archaic or vague that the police could probably arrest you simply for listening to the radio.

Many states are trying to repeal or amend monitoring laws. For instance, New Jersey and Kentucky have amended their laws, and ethical hackers in those states have little fear of prosecution. On the other hand, in California, monitoring cordless telephone calls is illegal. But the common-law landscape is ever-changing. The laws at any time are no more than a snapshot. Lawmakers pass new laws every day, while discarding or amending old ones. This is inevitable. So what can you do? You need to keep as informed as you can about your local laws to be completely safe.

In the meantime, the current monitoring laws break down into three categories:

✔ **Outright bans on monitoring by everyone:** Outright bans on monitoring exist in Florida, Indiana, Kentucky, Michigan, Minnesota, and New York. Usually, the laws make allowances for people like amateur radio operators, journalists, and law enforcement personnel.

✔ **Outright bans on monitoring by criminals:** South Dakota and Rhode Island prohibit criminals they list in the acts from having scanners.

✔ **Bans on illegal use of discovered information:** California, Minnesota, Nebraska, New Jersey, New York, Oklahoma, Vermont, and West Virginia fall in this category. In simple terms, their laws forbid the improper *use* (not the listening to) of information.

Let the master answer

A key doctrine in the law of agency is *respondeat superior,* which is Latin for "let the master answer." This doctrine provides that a principal (your employer) is responsible for the actions of his/her/its agent (you) in the "course of employment." Thus, when you, acting as an agent, connect to an access point while performing an ethical hack for your employer in the name of the employer, your employer has responsibility for your actions. So if you cause harm, the injured party may hold your employer accountable. Hey, enough legalese. We only play lawyers on TV.

Although North Dakota prohibits possession of a radio that will both transmit and receive police frequencies, other states do not appear currently to have laws restricting radio listening. To protect yourself, you should also familiarize yourself with the Electronic Communications Privacy Act (ECPA) and any state statutes that support the ECPA.

New Hampshire has an interesting approach to wardriving. House Bill 495 effectively legalizes many forms of wardriving. Legal eagles have interpreted this law to mean that any unsecured wireless network in the state of New Hampshire is fair game. Therefore, if wireless network administrators in New Hampshire want to prosecute people for hacking into their wireless net works, they need to have taken action to secure their networks. Not surprisingly, the Electronic Freedom Frontier (www.eff.org) feels that the New Hampshire law is a step in the right direction.

Warchalking, a seemingly innocuous activity, may actually run afoul of the law, and not just because warchalkers are tempting others to break laws. Your jurisdiction may have laws against graffiti or marking sidewalks.

Also, some states prohibit having or watching screens like TVs and computers while you're driving.

The one thing we, as "pseudo-legals," know is that there are no cut-and-dried answers to these legal questions. As with any questionable activity, subtle nuances rule the day. However, simply driving around a city searching for the existence of wireless networks, with no ulterior motives, is most likely legal. Courts haven't exhaustively tested the concept, but the common assumption is that simply discovering open wireless networks is legal.

But if you search for an open network in order to steal Internet access or commit computer crimes, then the wardriving you performed was done in a malicious manner, and most courts will treat it as such. What is certainly illegal is connecting to and using networks without the network owner's permission, what most people call "breaking into a network." Existing laws, which vary from state to state, already cover "cybercrime," such as using a neighbor's wireless network for downloading child pornography or broadcasting spam. As we mention in Chapter 2, charges against individuals so far have related to other crimes, such as violations of the CAN-SPAM Act of 2003, and not to the wardriving activities in and of themselves. In Toronto, for example, a man piggybacked on a wireless connection to download child pornography. The man was not charged with theft of services by the ISP, but with the more vile crime of possession of child pornography.

So you must demonstrate the required propriety when engaging in ethical hacking. Here are some simple rules to follow as an ethical hacker:

- ✔ Don't touch.
- ✔ Don't look.
- ✔ Don't play through.
- ✔ Don't violate your security policy.
- ✔ Don't operate beyond your agreement.
- ✔ Don't operate beyond the scope of the original work unless you have an official, signed-and-approved request.

In other words, you should not examine the contents of a network; add, delete, or change anything on the network; or use the network's resources. Follow the ecological advice given to hikers and "take only pictures and leave only footprints." Act responsibly, do the right thing, and don't trespass onto private Wi-Fi networks.

In conclusion, the law hasn't entirely caught up with networking as a whole, much less the peripheral issues of ethical hacking. So, obey the law as it exists today and don't engage in activity that is unethical. Public perception is extremely important.

Chapter 18

Ten Tips for Following Up after Your Testing

After you complete your wireless ethical-hacking tests, the work's not done. The vulnerabilities you've found are only a current snapshot of what's taking place, so time is of the essence. It's important to keep up your momentum and follow through to make sure you maximize the return on your efforts. By performing the steps in this chapter, you'll have a more secure wireless network — not only now, but also in the future.

Organize and Prioritize Your Results

During your ethical-hacking tests, you've likely amassed a large amount of vulnerability information and test data from your hacking tools. It's critical to comb through this information and organize your vulnerabilities into a readable and manageable format. The idea is to create a good roadmap for addressing these issues, both now and later.

Break your vulnerabilities down into groups similar to the breakdown of the various chapters of this book, such as default-settings weaknesses, wireless-client issues, encryption problems, and so on. Next, prioritize your vulnerabilities overall or within each category. You can use a rating system such as 1 for High (Must Address Now), 2 for Medium (Need to Address Soon), and 3 for Low (Should Address in the Future When Time/Resources Permit).

As you rate the problems, focus on both the *likelihood* that a vulnerability will be exploited and the *impact* to your organization if it is exploited. For example, WEP key-rotation or LEAP-authentication vulnerabilities would likely be a lower priority compared to radio signals leaking outside the building — or certain systems not running WEP encryption at all.

Prepare a Professional Report

The big deliverable for your project sponsor is your final report — outlining the wireless network security vulnerabilities you found, along with specific recommendations for fixing those problems. It's the final product that you and/or others will rely upon as you make security decisions in the future. Because various people — including company bigwigs — may have access to this report, make it look professional. This is important not just for the sake of readability — it also plays a role in whether you'll be asked to do this type of work for your project sponsor again in the future. Charts, tables, and other easy-to-refer-to forms of graphical data are especially nice. Just don't focus too much on style over substance.

Retest If Necessary

As you pull your test results together, you may discover some interesting or unexpected results that you may need to look into further. In addition, you may realize that you've overlooked a system and need to go back and retest. Don't fret — that's okay. Simply retest the systems and either integrate your new results with your existing data or add an addendum to your report that outlines your latest findings.

Obtain Sign-Off

It's important to get sign-off from your project sponsor: a written statement from the sponsor to acknowledge that your work is complete. If you'll be obtaining your boss's sign-off, doing so can be as informal as an e-mail (or even her signature on the cover page of the final report). Written acknowledgement is especially important for independent consultants: It helps you get paid for all your efforts! The bottom line is that you must be sure your sponsor agrees the work is complete — and is willing to say so in writing.

Plug the Holes You Find

This is a no-brainer. The whole reason you're performing ethical-hacking tests against your wireless systems is to make them more secure. Make sure that you or the responsible parties follow up and actually address the vulnerabilities you've found — especially the high- and medium-priority items.

Document the Lessons Learned

One eternal principle of network security that we've discovered over the years is that you have to learn from past experiences — and pass it on. The best way to do so is to document the information your testing has uncovered. Don't just report — recommend. If you make mistakes during your testing, discover a better practice than the one you normally use; if you simply want to make notes that'll streamline your ethical-hacking efforts in the future, then document it all — the methods, the specific tools used, the problems uncovered, the remedies. Simply pull out pen and paper or fire up your favorite word processor and create a document. Make sure you keep a hard copy and update periodically. You can document as soon as you discover something new or after your testing is completed — whatever works best for you. This information will be invaluable; time spent documenting the work makes the work more effective.

Repeat Your Tests

One of the downsides to ethical hacking is that all the tests you perform provide only a snapshot in time of your wireless network vulnerabilities. Changes in the system over time produce new vulnerabilities, and new threats are always emerging. If you truly want to keep your systems solid and secure into the future, you've got to repeat the tests we've outlined in this book over and over again. How often? Well, there is no one best schedule since everyone's needs are different. You may find that monthly, quarterly, bi-annual, or annual tests are best. If in doubt, err on the side of caution and test as often as time, money, and manpower permit.

Monitor Your Airwaves

Another part of your ongoing efforts should include the periodic monitoring of your airwaves for any changes. Look for additional wireless clients that have joined your network, network traffic, protocol utilization, and evidence of other wireless systems in close proximity to yours. A basic software tool such as Network Stumbler will work for small wireless networks, but for larger deployments, it would behoove you to use something more scalable and manageable such as a wireless-network analyzer or even a wireless and intrusion detection or prevention system.

Practice Using Your Wireless Tools

Throughout this book, we've covered an array of wireless-security testing tools. Many of these tools are complex and can take years to master. Well, why not start the mastery process now? Get your hands dirty by using your tools on an ongoing basis. Familiarize yourself with their functionality, especially those features that you've never used. This is especially important for a tool such as a wireless network analyzer. You might just find something that'll be of benefit when your formal testing time comes around. This learn-by-doing approach can be more help to your becoming an expert ethical hacker than anything else.

Keep Up with Wireless Security Issues

For the true pro, school is never really out. As with any worthy technology, 802.11 wireless-security issues are always popping up. The best way to stay unsurprised by them is to stay tuned in to your own favorite wireless-security resources (our favorites are the security resources we outline in this book's Appendix A). Knowledge is power, so make a point to keep up with what's going on in the wireless world.

Part V
Appendixes

The 5th Wave   By Rich Tennant

"Frankly, the idea of an entirely wireless future
scares me to death."

In this part . . .

We feel that this book is just the start of your journey. If you want to hone your proficiency in ethical hacking, then you'll need to use this book as a launch pad — and keep up with the field over the longer term. By the time this book comes to press, for example, it's a good bet someone will have released the next big cracker tool. You'll need to come up with a means of keeping abreast of the latest technology and tools.

To this end, we thought you would benefit from some links to organizations that specialize in wireless standards and some references to wireless user groups (WUGs). Find a group in your neighborhood and get involved. And, should you not know all the intricacies of wireless geek-speak, we have provided a list of acronyms and initialisms.

Appendix A
Wireless Hacking Resources

. . . .

In This Appendix

▶ Making contact with wireless organizations

▶ Finding local wireless user groups

▶ Shopping for wireless tools

. . . .

*W*ireless networking is evolving extremely fast. To keep your company and yourself current, you will need to keep up to date on developing standards and tools. This book gets you started, but learning is a life-long experience. We have listed some organizations and tools to help.

Certifications

We covered a lot about wireless and ethical hacking in this book. You may want to find out how knowledgeable you now are. The best way to do that is to take a certification test. Following are two organizations that certify individuals on this material.

> ✔ **Certified Ethical Hacker:** www.eccouncil.org/CEH.htm

> ✔ **Certified Wireless Network Professional Program:** www.cwnp.com

General Resources

The Internet is a valuable resource. However, using it is like trying to get a sip of water from a fire hose. So you need to damper the flow of information. We have found that the following sites provide useful information on wireless on a recurring basis. They also have free subscription mailing lists.

> ✔ **SearchMobileComputing.com:** www.searchmobilecomputing.com

> ✔ **SearchNetworking.com:** www.searchnetworking.com

> ✔ **SearchSecurity.com:** www.searchsecurity.com

Hacker Stuff

Sun Tzu in the "Art of War" writes that you must understand your enemy to defeat your enemy. Learning about your enemy is a good tactic. When you can put yourself in the mindset of your enemy then you can truly understand your enemy. There are many good "hacker" sites available to you. Following are several sites that will help you understand crackers.

- **2600 — The Hacker Quarterly magazine:** www.2600.com
- *Computer Underground Digest:* www.soci.niu.edu/~cudigest
- **Hacker t-shirts, equipment, and other trinkets:** www.thinkgeek.com
- **Honeypots: Tracking Hackers:** www.tracking-hackers.com
- **The Online Hacker Jargon File:** www.jargon.8hz.com
- **PHRACK:** www.phrack.org

Wireless Organizations

There are two wireless organizations that you need to acquaint yourself with. These are the IEEE and the Wi-Fi Alliance. The former concerns itself with setting standards for wireless, and the latter certifies that WLAN equipment meets the standards set by the former.

Institute of Electrical and Electronics Engineers (IEEE): www.ieee.org

In this book we mention the pertinent wireless standards: 802.11, 802.11a, 802.11b, 802.11g, and 802.11i. These standards are all the creations of the Institute of Electrical and Electronic Engineers (IEEE). The IEEE leads the way in developing open standards for Wireless Local Area Networks (Wireless LANs), Wireless Personal Area Networks (Wireless PANs), and Wireless Metropolitan Area Networks (Wireless MANs). You can compare and contrast the 802.11 wireless standards for "over the air" to the 802.3 Ethernet standards for "over the wire."

Wi-Fi Alliance (formerly WECA): www.wifialliance.com

Formed in 1999, the Wi-Fi Alliance is a nonprofit association that certifies the interoperability of wireless Local Area Network products that are based on IEEE 802.11 specifications. The Wi-Fi Alliance has over 200 member companies from around the world — and has certified over 1,000 devices. All the equipment used in the making of this book (for example) was tested and certified by the Wi-Fi Alliance — and without animal testing!

Local Wireless Groups

Should you really want to get serious about wireless ethical hacking, you'll need to immerse yourself in the culture. Hook up with other wireless aficionados, who can turn you on to new tools and point you to useful whitepapers and other resources. Wireless grassroots organizations are springing up like crabgrass across the world. You can meet like-minded wireless buffs and do some networking — the social kind. Here is a sampling of wireless user groups:

- **Air-Stream, Adelaide, SA, AU:** www.air-stream.org/tiki-custom_home.php
- **AirShare, San Diego, CA, US:** www.airshare.org/
- **Albany Wireless User Group, Albany, NY, US:** http://community.albanywifi.com/index.php
- **Austin Wireless, Austin, TX, US:** www.austinwireless.net
- **Barcelona Wireless, Barcelona, Cataluña, ES:** http://barcelonawireless.net/
- **Bay Area Wireless Users Group (BAWUG), Bay Area, CA, US:** www.bawug.org
- **BC Wireless, Vancouver, Vancouver Island and Prince Rupert, BC, CA:** http://bcwireless.net/
- **Capital Area Wireless Network, Northern Virginia, VA, US:** www.cawnet.org/
- **Consume, London, England, UK:** www.consume.net/

- ✔ **Corkwireless, Cork, Cork County, IE:** www.corkwireless.com/
- ✔ **Georgia Wireless User Group, Atlanta, GA, US:** www.gawug.com
- ✔ **Green Bay Professional Packet Radio, Green Bay, WI, US:** www.qsl.net/n9zia/
- ✔ **Houston Wireless, Houston, TX, US:** www.houstonwireless.org
- ✔ **IrishWAN, IE:** www.irishwan.org/
- ✔ **Longmount Community Wireless Project, Longmount, CO, US:** http://long-wire.net/
- ✔ **Madrid Wireless, Madrid, Madrid, ES:** http://madridwireless.net/
- ✔ **Marin Unwired, Marin County, CA, US:** www.digiville.com/wifi-marin/index.htm
- ✔ **NoCatNet, Sonoma County, CA, US:** http://nocat.net
- ✔ **NYCWireless, New York City, NY, US:** http://nycwireless.net
- ✔ **NZ Wireless, Auckland, NZ:** www.nzwireless.org/
- ✔ **Orange County California Wireless Users Group, Brea, CA, US:** www.occalwug.org/
- ✔ **Personal Telco, Portland, OR, US:** www.personaltelco.net
- ✔ **Rooftops, Boston/Cambridge, MA, US:** http://rooftops.media.mit.edu/
- ✔ **Salt Lake Area Wireless Users Group (SLWUG), Salt Lake City, UT, US:** www.saltlakewireless.net/
- ✔ **San Diego Wireless Users Group, San Diego, CA, US:** www.sdwug.org
- ✔ **Seattle Wireless, Seattle, WA, US:** www.seattlewireless.net
- ✔ **Southern California Wireless Users Group, Southern California, CA, US:** www.socalwug.org
- ✔ **StockholmOpen.net, Stockholm, SE:** www.stockholmopen.net/index.php
- ✔ **The Toronto Wireless User Group (TorWUG), Toronto, ON, CA:** www.torwug.org/
- ✔ **Tri-Valley Wireless Users Group, US:** www.tvwug.org
- ✔ **Xnet Wireless, Mornington, AU:** www.x.net.au/
- ✔ **WiFi Ecademy, London, England, UK:** www.wifi.ecademy.com/
- ✔ **Wireless Technology Forum, Atlanta, GA, US:** www.wirelesstechnologyforum.com
- ✔ **Wireless France, FR:** www.wireless-fr.org/spip/

If you can't find your location from this list, then try the following sites to find a user group near you:

- ✔ `www.practicallynetworked.com/tools/wireless_articles_ community.htm`
- ✔ `www.wirelessanarchy.com/#Community%20Groups`
- ✔ `www.personaltelco.net/index.cgi/WirelessCommunities`

Security Awareness and Training

You may find that getting management and staff to pay attention to information security is at best a difficult task. You are not alone. Fortunately the following companies can help you get the message across in your organization.

- ✔ **Greenidea, Inc. Visible Statement:** `www.greenidea.com`
- ✔ **The Security Awareness Company:** `www.thesecurityawareness company.com`
- ✔ **Security Awareness, Inc. Awareness Resources:** `www.security awareness.com`
- ✔ **U.S. Security Awareness:** `www.ussecurityawareness.org`

Wireless Tools

Throughout the book, we have described many tools — showing where to get them, classifying, and summarizing them. If you are just starting out, the tools listed here make a nice shopping list. If you are getting married, you can register at `hackersrus.com`. Ethical-hacking tools also make great anniversary gifts for those two-hacker households.

General tools

We have grouped tools into specific categories. But some of them defied categorization. But rather than lose these excellent tools you can use, we offer the following list:

- ✔ **BLADE Software IDS Informer:** `www.bladesoftware.net`
- ✔ **Foundstone SiteDigger Google query tool:** `www.foundstone.com/ resources/freetools.htm`

✔ **MAC-address-vendor lookup:** `http://coffer.com/mac_find`

✔ **SMAC MAC-address editor for Windows:** `www.klcconsulting.net/smac/`

✔ **WiGLE database:** `www.wigle.net/gps/gps/GPSDB/query/`

✔ **WiFimaps:** `www.wifimaps.com`

Vulnerability databases

You will need to understand the vulnerabilities associated with your particular hardware and software. During the planning process, you will use this information to determine the exact tests to perform. Following are some well-known vulnerability database sites.

✔ **US-CERT Vulnerability Notes Database:** `www.kb.cert.org/vuls`

✔ **NIST ICAT Metabase:** `http://icat.nist.gov/icat.cfm`

✔ **Common Vulnerabilities and Exposures:** `http://cve.mitre.org/cve`

Linux distributions

Since many wireless testing tools only run on UNIX, Linux or BSD, you will need to become familiar with one of these platforms. You can purchase a commercial product like SuSe or Red Hat Linux, but this is overkill for our purposes. So instead use one of the following freeware Linux distributions.

✔ **Auditor:** `http://new.remote-exploit.org/index.php/Auditor_main`

✔ **Cool Linux CD:** `http://sourceforge.net/project/showfiles.php?group_id=55396&release_id=123430`

✔ **DSL (Damn Small Linux):** `www.damnsmalllinux.org/`

✔ **GNU/Debian Linux:** `www.debian.org/`

✔ **KNOPPIX:** `www.knoppix.net/get.php`

✔ **SLAX:** `http://slax.linux-live.org/`

✔ **WarLinux:** `http://sourceforge.net/projects/warlinux/`

Software emulators

If you want to run more than one operating system at a time on the same hardware or want to paste from one operating system to another, then you will want to consider a software emulation product. Following are some of the better-known products.

- **Bochs:** http://bochs.sourceforge.net/
- **Cygwin:** http://cygwin.com/
- **DOSEMU:** www.dosemu.org/
- **Microsoft Virtual PC:** www.microsoft.com/mac/products/ virtualpc/virtualpc.aspx?pid=virtualpc
- **Plex86:** http://savannah.nongnu.org/projects/plex86/
- **Vmware:** www.vmware.com/
- **WINE:** www.winehq.com/
- **Win4lin:** www.netraverse.com/

RF prediction software

RF prediction software helps you simulate the radiation pattern of an access point without having to physically install one. So as a tester you use the same software to predict where you may find a signal. Following are three such software programs.

- **Airespace:** www.airespace.com/products/AS_ACS_location_ tracking.php
- **Alcatel:** www.ind.alcatel.com/products/index.cfm?cnt= omnivista_acs_locationtrack
- **Radioplan:** www.electronicstalk.com/news/rop/rop100.html

RF monitoring

You can use software to monitor signal strength and bit error rate. Of course, tools like Kismet or NetStumbler give you signal strength, but they don't do it as well as the following tools.

- ✔ **aphunter:** www.math.ucla.edu/~jimc/mathnet_d/download.html
- ✔ **E-Wireless:** www.bitshift.org/wireless.shtml
- ✔ **Gkrellm wireless plug-in:** http://gkrellm.luon.net/gkrellm wireless.phtml
- ✔ **Gnome Wireless Applet:** http://freshmeat.net/projects/ gwifiapplet/
- ✔ **Gtk-Womitor:** www.zevv.nl/wmifinfo/
- ✔ **GWireless:** http://gwifiapplet.sourceforge.net/
- ✔ **Kifi:** http://kifi.staticmethod.net/
- ✔ **KOrinoco:** http://korinoco.sourceforge.net/
- ✔ **KWaveControl:** http://kwavecontrol.sourceforge.net/
- ✔ **KWiFiManager:** http://kwifimanager.sourceforge.net/
- ✔ **Linux Wireless Extensions:** http://pcmciacs.sourceforge.net/ ftp/contrib/
- ✔ **Mobydik.tk:** www.cavone.com/services/mobydik_tk.aspx
- ✔ **NetworkControl:** www.arachnoid.com/NetworkControl/index.html
- ✔ **NetworkManager:** http://people.redhat.com/dcbw/Network Manager/
- ✔ **Qwireless:** www.uv-ac.de/qwireless/
- ✔ **Wavemon:** www.janmorgenstern.de/wavemon-current.tar.gz
- ✔ **WaveSelect:** www.kde-apps.org/content/show.php?content=19152
- ✔ **Wimon:** http://imil.net/wimon/
- ✔ **Wmap:** www.datenspuren.org/wmap
- ✔ **wmifinfo:** www.zevv.nl/wmifinfo/
- ✔ **WMWave:** www.schuermann.org/~dockapps/
- ✔ **WmWiFi:** http://wmwifi.digitalssg.net/?sec=1
- ✔ **Wscan:** www.handhelds.org/download/packages/wscan/
- ✔ **wvlanmon:** http://file.wankota.org/program/linux/wavelan/
- ✔ **XNetworkStrength:** http://gabriel.bigdam.net/home/ xnetstrength/
- ✔ **xosview:** http://open-linux.de/index.html.en

Antennae

You can spend a lot of money on an antenna. However, you need not spend all that money. You can build one yourself or acquire one for a pretty reasonable sum. Following are three sites to help you acquire an economical antenna for your ethical-hacking work.

- ✔ **Cantenna:** www.cantenna.com
- ✔ **Hugh Pepper's cantennas, pigtails, and supplies:** http://home.comast.net/~hughpep
- ✔ **Making a wireless antenna from a Pringles can:** www.oreilly net.com/cs/weblog/view/wlg/448

You can find a very good reference page for antennae at www.wardrive.net/general/antenna.

Wardriving

A very useful tool for your wireless ethical-hacking kit is a wardriving or network discovery program. Fortunately for you, there is an overabundance of tools as the following list shows.

- ✔ **Aerosol:** www.sec33.com/sniph/aerosol.php
- ✔ **AirMagnet:** www.airmagnet.com/products/index.htm
- ✔ **AiroPeek:** www.wildpackets.com/products/airopeek
- ✔ **Airscanner:** www.snapfiles.com/get/pocketpc/airscanner.html
- ✔ **AP Scanner:** www.macupdate.com/info.php/id/5726
- ✔ **AP Radar:** http://apradar.sourceforge.net
- ✔ **Apsniff:** www.monolith81.de/mirrors/index.php?path=apsniff/
- ✔ **BSD-Airtools:** www.dachb0den.com/projects/bsd-air tools.html
- ✔ **dstumbler:** www.dachb0den.com/projects/dstumbler.html
- ✔ **gtk-scanner:** http://sourceforge.net/projects/wavelan-tools
- ✔ **gWireless:** http://gwifiapplet.sourceforge.net/
- ✔ **iStumbler:** http://istumbler.net/
- ✔ **KisMAC:** www.binaervarianz.de/projekte/programmieren/kismac/

- ✔ **Kismet:** www.kismetwireless.net

- ✔ **MacStumbler:** www.macstumbler.com/

- ✔ **MiniStumbler:** www.netstumbler.com/downloads/

- ✔ **Mognet:** www.l0t3k.net/tools/Wireless/Mognet-1.16.tar.gz

- ✔ **NetChaser:** www.bitsnbolts.com

- ✔ **Network Stumbler:** www.netstumbler.com/downloads

- ✔ **perlskan:** http://sourceforge.net/projects/wavelan-tools

- ✔ **PocketWarrior:** www.pocketwarrior.org/

- ✔ **pocketWinc:** www.cirond.com/pocketwinc.php

- ✔ **Prismstumbler:** http://prismstumbler.sourceforge.net

- ✔ **Sniff-em:** www.sniff-em.com

- ✔ **Sniffer Wireless:** www.networkgeneral.com/

- ✔ **StumbVerter:** www.michiganwireless.org/tools/Stumbverter/

- ✔ **THC-Scan:** www.thc.org/releases.php?q=scan

- ✔ **THC-WarDrive:** www.thc.org/releases.php?q=wardrive

- ✔ **WarGlue:** www.lostboxen.net/warglue/

- ✔ **WarKizniz:** www.michiganwireless.org/tools/WarKizNiz/

- ✔ **Wellenreiter:** www.wellenreiter.net/

- ✔ **Wi-Scan:** www.michiganwireless.org/tools/wi-scan/

- ✔ **WiStumbler:** www.gongon.com/persons/iseki/wistumbler/index.html

- ✔ **Wireless Security Auditor:** www.research.ibm.com/gsal/wsa/

- ✔ **Wlandump:** www.guerrilla.net/gnet_linux_software.html

Wireless IDS/IPS vendors

Wireless IDS/IPS products are necessary whether you support wireless networking or not in your organization. If you do support wireless, then you need a tool to protect your network. If you don't have wireless, then you need a tool to ensure you don't. Following are some IDS/IPS products.

- ✔ **AirDefense:** www.airdefense.net

- ✔ **AirMagnet:** www.airmagnet.com

- **BlueSocket:** www.bluesocket.com
- **ManageEngine:** http://origin.manageengine.adventnet.com/products/wifi-manager
- **NetMotion Wireless:** www.netmotionwireless.com
- **Red-Detect:** www.red-m.com/Products/Red-Detect
- **Senforce Wi-Fi Security:** www.senforce.com/entwirelessecur.htm
- **Vigilant Minds:** www.vigilantminds.com
- **WiFi Manager:** http://manageengine.adventnet.com/products/wifi-manager/index.html

Wireless sniffers

You know that old saw: a picture is worth a thousand words. Well, the message from the saw applies to ethical hacking. Show someone his password that you captured because it wasn't encrypted, and he gets it. Following are some packet capture tools.

- **AirMagnet:** www.airmagnet.com/
- **AiroPeek:** www.wildpackets.com/products/airopeek
- **AirScanner Mobile Sniffer:** http://airscanner.com/downloads/sniffer/sniffer.html
- **AirTraf:** http://airtraf.sourceforge.net/
- **Capsa:** www.colasoft.com/products/capsa/index.php?id=75430g
- **CENiffer:** www.epiphan.com/products_ceniffer.html
- **CommView for WiFi:** www.tamos.com/products/commview/
- **ethereal:** www.ethereal.com
- **Gulpit:** www.crak.com/gulpit.htm
- **KisMAC:** www.binaervarianz.de/projekte/programmieren/kismac/
- **Kismet:** www.kismetwireless.net/
- **LANfielder:** www.wirelessvalley.com/
- **LinkFerret:** www.baseband.com/
- **Mognet:** www.l0t3k.net/tools/Wireless/Mognet-1.16.tar.gz

  ✔ **ngrep:** `www.remoteassessment.com/?op=pub_archive_search&query=wireless`

  ✔ **Observer:** `www.networkinstruments.com/`

  ✔ **Packetyzer:** `www.networkchemistry.com/`

  ✔ **Sniffer Netasyst:** `www.sniffer-netasyst.com/`

  ✔ **Sniffer Wireless:** `www.networkgeneral.com/Products_details.aspx?PrdId=20046178370181`

WEP/WPA cracking

If we had a dollar for every time someone said she's OK because she uses WEP or WPA, we would retire to a nice island in the Caribbean. The following tools should show them that they are not OK.

  ✔ **Aircrack:** `www.cr0.net:8040/code/network/`

  ✔ **AirSnort:** `http://sourceforge.net/projects/airsnort/`

  ✔ **Destumbler:** `http://sourceforge.net/projects/destumbler`

  ✔ **Dwepcrack:** `www.e.kth.se/~pvz/wifi/`

  ✔ **jc-wepcracker:** `www.astalavista.com/?section=dir&cmd=file&id=3316`

  ✔ **Lucent Orinoco Registry Encryption/Decryption program:** `www.cqure.net/tools.jsp?id=3`

  ✔ **WepAttack:** `http://wepattack.sourceforge.net/`

  ✔ **WEPcrack:** `http://sourceforge.net/projects/wepcrack/`

  ✔ **WEPWedgie:** `http://sourceforge.net/projects/wepwedgie/`

  ✔ **WepLab:** `http://weplab.sourceforge.net/`

  ✔ **WinAirSnort:** `www.nwp.nevillon.org/attack.html`

  ✔ **WPA Cracker:** `www.tinypeap.com/page8.html`

Cracking passwords

There are tools that will grab packets, look for passwords, and provide them to you. Following are some of these very desirable tools.

  ✔ **Cain & Abel:** `www.oxid.it/cain.html`

  ✔ **Dsniff:** `www.monkey.org/~dugsong/dsniff/`

✔ **Dsniff (Windows port):** www.datanerds.net/~mike/dsniff.html

✔ **Dsniff (MacOS X port):** http://blafasel.org/~floh/ports/dsniff-2.3.osx.tgz

Crack only passwords that you have the authority to crack. Cracking other passwords could end you up in jail.

Dictionary files and word lists

Most password crackers take a list of words or a dictionary and encrypt the words and then compare them to the password file. So you need to get different dictionaries or wordlists. Following are five good sources for dictionaries and wordlists.

✔ **CERIAS Dictionaries and Wordlists:** ftp://ftp.cerias.purdue.edu/pub/dict

✔ **Default vendor passwords:** www.cirt.net/cgi-bin/passwd.pl

✔ **Outpost9 Wordlists:** www.outpost9.com/files/WordLists.html

✔ **PacketStorm Wordlists:** http://packetstormsecurity.nl/Crackers/wordlists

✔ **University of Oxford Dictionaries and Wordlists:** ftp://ftp.ox.ac.uk/pub/wordlists

Gathering IP addresses and SSIDs

Many wireless security books recommend that you turn off SSID broadcasting as a control. However, you can use one of the following programs to get the SSID even when they do.

✔ **air-jack:** http://sourceforge.net/projects/airjack/

✔ **Arping:** www.habets.pp.se/synscan/programs.php?prog=arping

✔ **essid_jack:** http://sourceforge.net/projects/airjack/

✔ **pong:** http://mobileaccess.de/wlan/index.html?go=technik&sid=

✔ **SSIDsniff:** www.bastard.net/~kos/wifi/ssidsniff-0.40.tar.gz

LEAP crackers

EAP is touted as the solution to the WEP authentication problem. However EAP has its own problems. Following are three tools you can use to crack LEAP.

- **anwrap:** http://packetstormsecurity.nl/cisco/anwrap.pl
- **asleap:** http://asleap.sourceforge.net/
- **THC-LEAPcracker:** http://thc.org/releases.php?s=4&q=&o=

Network mapping

After you connect to an access point, you will want to map the network. You will want to know how many servers you can find and what operating system the server is running. Following are some tools to help you map your network.

- **Cheops:** www.marko.net/cheops/
- **Cheops-ng:** http://cheops-ng.sourceforge.net
- **SNMPUTIL.EXE:** www.microsoft.com
- **Snmpwalk:** www.trinux.org
- **Solarwinds Standard Edition Version:** www.solarwinds.net
- **WhatsUp Gold:** www.ipswitch.com/products/network-management.html

Network scanners

Network scanners help you identify applications running on the systems on your network. You may find these applications on servers and network devices alike. Following are some that we have used.

- **fping:** www.fping.com
- **GFI LANguard Network Security Scanner:** www.gfi.com/lannetscan
- **nessus:** www.nessus.org
- **nmap:** www.insecure.org/nmap
- **QualysGuard:** www.qualys.com
- **SoftPerfect Network Scanner:** www.softperfect.com/products/networkscanner
- **SuperScan:** www.foundstone.com/resources/proddesc/superscan.htm

Appendix B
Glossary of Acronyms

3DES: Triple Data Encryption Standard

ACK: Acknowledge

ACL: Access Control List

AES: Advanced Encryption Standard

AES-CCMP: ES-Counter Mode CBC-MAC Protocol

AES-WRAP: ES-Wireless Robust Authenticated Protocol

AH: Authentication Header

AP: Access Point

BBWA: Broadband Wireless Access

BER: Bit Error Rate

BSS: Basic Service Set

BSSID: Basic Service Set Identifier

CCK: Complimentary Code Keying

CF: Compact Flash

CHAP: Challenge/Handshake Authentication Protocol

CRC: Cyclic Redundancy Check

CSMA/CA: Carrier Sense Multiple Access/Collision Avoidance

CTS: Clear to Send

DB: Decibel

DBm: Decibel per milliwatt

DBPSK: Differential Binary Phase Shifting Key

DCF: Distributed Coordination Function

DDoS: Distributed Denial of Service

DES: Data Encryption Standard

DHCP: Dynamic Host Configuration Protocol

DiGLE: Delphi imaging Geographic Lookup Engine

DMZ: De-Militarized Zone

DoS: Denial of Service

DQPSK: Differential Quadrature Phase Shifting Key

DSSS: Direct Sequence Spread Spectrum

EAP: Extensible Authentication Protocol

EAP-TLS: AP-Transport Layer Security

EAP-TTLS: EAP-Tunneled Transport Layer Security

EAPOL: EAP Over LANs

ESP: Encapsulating Security Protocol

ESS: Extended Service Set

ESSID: Extended Service Set Identifier

FCC: Federal Communications Commission

FH: Frequency Hopping

FHSS: Frequency Hopping Spread Spectrum

FIN: Finish

GFSK: Gaussian Phase Shifting Key

GHz: Gigahertz

GPS: Global Positioning System

GSM: Global System for Mobile Communications

HR/DSSS: High-Rate Direct-Sequence Spread Spectrum

HTTP: Hypertext Transfer Protocol

IAPP: Inter-Access Point Protocol

IBSS: Independent Basic Service Set

ICAT: Internet Categorization of Attack Toolkit

ICV: Integrity Check Value

IDS: Intrusion Detection System

IEEE: Institute of Electrical and Electronics Engineers

IETF: Internet Engineering Task Force

IKE: Internet Key Exchange

IP: Internet Protocol

IPS: Intrusion Prevention System

Ipsec: Internet Protocol Security

ISM: Industrial, Scientific, and Medical

ISO: International Organization for Standardization

IV: Initialization Vector

JiGLE: Java-imaging Geographic Lookup Engine

Kbps: Kilobits per second

KHz: Kilohertz

L2TP: Layer 2 Tunneling Protocol

LAN: Local Area Network

LBT: Listen Before Talking

LDAP: Lightweight Directory Access Protocol

LEAP: Lightweight EAP

LLC: Logical Link Control

LOS: Line of Sight

MAC: Media Access Control

Mbps: Megabits per second

MD5: Message Digest 5

MHz: Megahertz

MIB: Management Information Base

MIC: Message Integrity Check

mW: Milliwatt

MIMO: Multiple-In/Multiple-Out

MITM: Man-in-the-Middle; Monkey-in-the-Middle

NIC: Network Interface Card

OFDM: Orthogonal Frequency Division Multiplexing

PAP: Password Authentication Protocol

PCF: Point Coordination Function

PCMCIA: Personal Computer Memory Card International Association

PDA: Personal Digital Assistant

PEAP: Protected EAP

PED: Personal Electronic Device

PKI: Public-Key Infrastructure

PoE: Power over Ethernet

PPTP: Point-to-Point Tunneling Protocol

PS-Poll: Power Save Poll

PSK: Pre-Shared Key

QAM: Quadrature Amplitude Modulation

RADIUS: Remote Authentication Dial-in User Service

RBAC: Role-Based Access Control

RC4: Ron's Code 4

RF: Radio Frequency

RF LOS: Radio Frequency Line of Sight

RSA: Rivest-Shamir-Adelman

RSN: Robust Security Networks

RTS: Request to Send

SME: Small-to-Medium Enterprise

SNMP: Simple Network Management Protocol

SNR: Signal-to-Noise Ratio

SOHO: Small Office Home Office

SSH: Secure Shell

SSID: Service Set Identifier

SSL: Secure Sockets Layer

SYN: Synchronize

TCP: Transmission Control Protocol

TCP/IP: Transmission Control Protocol/Internet Protocol

THC: The Hacker's Choice

TKIP: Temporal Key Integrity Protocol

TLS: Transport Layer Security

UDP: User Datagram Protocol

USB: Universal Serial Bus

VPN: Virtual Private Network

WAP: Wireless Application Protocol

WEP: Wired Equivalent Privacy

Wi-Fi: Wireless Fidelity

WIDS: Wireless Intrusion Detection System

WiGLE: Wireless Geographic Logging Engine

WISP: Wireless Internet Service Provider

WLAN: Wireless Local Area Network

WMM: Wi-Fi Multimedia

WPA: Wi-Fi Protected Access

Index

● *F* ●

• K •

BUSINESS, CAREERS & PERSONAL FINANCE

0-7645-5307-0

0-7645-5331-3 *†

Also available:

- Accounting For Dummies †
  0-7645-5314-3
- Business Plans Kit For Dummies †
  0-7645-5365-8
- Cover Letters For Dummies
  0-7645-5224-4
- Frugal Living For Dummies
  0-7645-5403-4
- Leadership For Dummies
  0-7645-5176-0
- Managing For Dummies
  0-7645-1771-6

- Marketing For Dummies
  0-7645-5600-2
- Personal Finance For Dummies *
  0-7645-2590-5
- Project Management For Dummies
  0-7645-5283-X
- Resumes For Dummies †
  0-7645-5471-9
- Selling For Dummies
  0-7645-5363-1
- Small Business Kit For Dummies *†
  0-7645-5093-4

HOME & BUSINESS COMPUTER BASICS

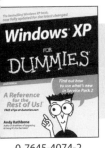

0-7645-4074-2

0-7645-3758-X

Also available:

- ACT! 6 For Dummies
  0-7645-2645-6
- iLife '04 All-in-One Desk Reference
  For Dummies
  0-7645-7347-0
- iPAQ For Dummies
  0-7645-6769-1
- Mac OS X Panther Timesaving
  Techniques For Dummies
  0-7645-5812-9
- Macs For Dummies
  0-7645-5656-8

- Microsoft Money 2004 For Dummies
  0-7645-4195-1
- Office 2003 All-in-One Desk Reference
  For Dummies
  0-7645-3883-7
- Outlook 2003 For Dummies
  0-7645-3759-8
- PCs For Dummies
  0-7645-4074-2
- TiVo For Dummies
  0-7645-6923-6
- Upgrading and Fixing PCs For Dummies
  0-7645-1665-5
- Windows XP Timesaving Techniques
  For Dummies
  0-7645-3748-2

FOOD, HOME, GARDEN, HOBBIES, MUSIC & PETS

0-7645-5295-3

0-7645-5232-5

Also available:

- Bass Guitar For Dummies
  0-7645-2487-9
- Diabetes Cookbook For Dummies
  0-7645-5230-9
- Gardening For Dummies *
  0-7645-5130-2
- Guitar For Dummies
  0-7645-5106-X
- Holiday Decorating For Dummies
  0-7645-2570-0
- Home Improvement All-in-One
  For Dummies
  0-7645-5680-0

- Knitting For Dummies
  0-7645-5395-X
- Piano For Dummies
  0-7645-5105-1
- Puppies For Dummies
  0-7645-5255-4
- Scrapbooking For Dummies
  0-7645-7208-3
- Senior Dogs For Dummies
  0-7645-5818-8
- Singing For Dummies
  0-7645-2475-5
- 30-Minute Meals For Dummies
  0-7645-2589-1

INTERNET & DIGITAL MEDIA

0-7645-1664-7

0-7645-6924-4

Also available:

- 2005 Online Shopping Directory
  For Dummies
  0-7645-7495-7
- CD & DVD Recording For Dummies
  0-7645-5956-7
- eBay For Dummies
  0-7645-5654-1
- Fighting Spam For Dummies
  0-7645-5965-6
- Genealogy Online For Dummies
  0-7645-5964-8
- Google For Dummies
  0-7645-4420-9

- Home Recording For Musicians
  For Dummies
  0-7645-1634-5
- The Internet For Dummies
  0-7645-4173-0
- iPod & iTunes For Dummies
  0-7645-7772-7
- Preventing Identity Theft For Dummies
  0-7645-7336-5
- Pro Tools All-in-One Desk Reference
  For Dummies
  0-7645-5714-9
- Roxio Easy Media Creator For Dummies
  0-7645-7131-1

Available wherever books are sold. For more information or to order direct: U.S. customers visit www.dummies.com or call 1-877-762-2974.
U.K. customers visit www.wileyeurope.com or call 0800 243407. Canadian customers visit www.wiley.ca or call 1-800-567-4797.

 **WILEY**

SPORTS, FITNESS, PARENTING, RELIGION & SPIRITUALITY

0-7645-5146-9

0-7645-5418-2

Also available:

- Adoption For Dummies
  0-7645-5488-3
- Basketball For Dummies
  0-7645-5248-1
- The Bible For Dummies
  0-7645-5296-1
- Buddhism For Dummies
  0-7645-5359-3
- Catholicism For Dummies
  0-7645-5391-7
- Hockey For Dummies
  0-7645-5228-7

- Judaism For Dummies
  0-7645-5299-6
- Martial Arts For Dummies
  0-7645-5358-5
- Pilates For Dummies
  0-7645-5397-6
- Religion For Dummies
  0-7645-5264-3
- Teaching Kids to Read For Dummies
  0-7645-4043-2
- Weight Training For Dummies
  0-7645-5168-X
- Yoga For Dummies
  0-7645-5117-5

TRAVEL

0-7645-5438-7

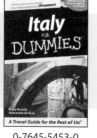

0-7645-5453-0

Also available:

- Alaska For Dummies
  0-7645-1761-9
- Arizona For Dummies
  0-7645-6938-4
- Cancún and the Yucatán For Dummies
  0-7645-2437-2
- Cruise Vacations For Dummies
  0-7645-6941-4
- Europe For Dummies
  0-7645-5456-5
- Ireland For Dummies
  0-7645-5455-7

- Las Vegas For Dummies
  0-7645-5448-4
- London For Dummies
  0-7645-4277-X
- New York City For Dummies
  0-7645-6945-7
- Paris For Dummies
  0-7645-5494-8
- RV Vacations For Dummies
  0-7645-5443-3
- Walt Disney World & Orlando For Dummies
  0-7645-6943-0

GRAPHICS, DESIGN & WEB DEVELOPMENT

0-7645-4345-8

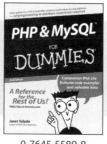

0-7645-5589-8

Also available:

- Adobe Acrobat 6 PDF For Dummies
  0-7645-3760-1
- Building a Web Site For Dummies
  0-7645-7144-3
- Dreamweaver MX 2004 For Dummies
  0-7645-4342-3
- FrontPage 2003 For Dummies
  0-7645-3882-9
- HTML 4 For Dummies
  0-7645-1995-6
- Illustrator CS For Dummies
  0-7645-4084-X

- Macromedia Flash MX 2004 For Dummies
  0-7645-4358-X
- Photoshop 7 All-in-One Desk
  Reference For Dummies
  0-7645-1667-1
- Photoshop CS Timesaving Techniques
  For Dummies
  0-7645-6782-9
- PHP 5 For Dummies
  0-7645-4166-8
- PowerPoint 2003 For Dummies
  0-7645-3908-6
- QuarkXPress 6 For Dummies
  0-7645-2593-X

NETWORKING, SECURITY, PROGRAMMING & DATABASES

0-7645-6852-3

0-7645-5784-X

Also available:

- A+ Certification For Dummies
  0-7645-4187-0
- Access 2003 All-in-One Desk
  Reference For Dummies
  0-7645-3988-4
- Beginning Programming For Dummies
  0-7645-4997-9
- C For Dummies
  0-7645-7068-4
- Firewalls For Dummies
  0-7645-4048-3
- Home Networking For Dummies
  0-7645-42796

- Network Security For Dummies
  0-7645-1679-5
- Networking For Dummies
  0-7645-1677-9
- TCP/IP For Dummies
  0-7645-1760-0
- VBA For Dummies
  0-7645-3989-2
- Wireless All In-One Desk Reference
  For Dummies
  0-7645-7496-5
- Wireless Home Networking For Dummies
  0-7645-3910-8

ALTH & SELF-HELP

0-7645-6820-5 *†

0-7645-2566-2

Also available:

- Alzheimer's For Dummies
  0-7645-3899-3
- Asthma For Dummies
  0-7645-4233-8
- Controlling Cholesterol For Dummies
  0-7645-5440-9
- Depression For Dummies
  0-7645-3900-0
- Dieting For Dummies
  0-7645-4149-8
- Fertility For Dummies
  0-7645-2549-2
- Fibromyalgia For Dummies
  0-7645-5441-7
- Improving Your Memory For Dummies
  0-7645-5435-2
- Pregnancy For Dummies †
  0-7645-4483-7
- Quitting Smoking For Dummies
  0-7645-2629-4
- Relationships For Dummies
  0-7645-5384-4
- Thyroid For Dummies
  0-7645-5385-2

DUCATION, HISTORY, REFERENCE & TEST PREPARATION

0-7645-5194-9

0-7645-4186-2

Also available:

- Algebra For Dummies
  0-7645-5325-9
- British History For Dummies
  0-7645-7021-8
- Calculus For Dummies
  0-7645-2498-4
- English Grammar For Dummies
  0-7645-5322-4
- Forensics For Dummies
  0-7645-5580-4
- The GMAT For Dummies
  0-7645-5251-1
- Inglés Para Dummies
  0-7645-5427-1
- Italian For Dummies
  0-7645-5196-5
- Latin For Dummies
  0-7645-5431-X
- Lewis & Clark For Dummies
  0-7645-2545-X
- Research Papers For Dummies
  0-7645-5426-3
- The SAT I For Dummies
  0-7645-7193-1
- Science Fair Projects For Dummies
  0-7645-5460-3
- U.S. History For Dummies
  0-7645-5249-X

Get smart @ dummies.com®

- **Find a full list of Dummies titles**
- **Look into loads of FREE on-site articles**
- **Sign up for FREE eTips e-mailed to you weekly**
- **See what other products carry the Dummies name**
- **Shop directly from the Dummies bookstore**
- **Enter to win new prizes every month!**

Separate Canadian edition also available
Separate U.K. edition also available

ailable wherever books are sold. For more information or to order direct: U.S. customers visit www.dummies.com or call 1-877-762-2974.
K. customers visit www.wileyeurope.com or call 0800 243407. Canadian customers visit www.wiley.ca or call 1-800-567-4797.